the universe

REVEALED

the universe REVEALED

General Editor **Pam Spence**

CAMBRIDGE
UNIVERSITY PRESS

To Jim, who found the joy of astronomy late in life,
but who gave so much.

PUBLISHED BY THE PRESS SYNDICATE
OF THE UNIVERSITY OF CAMBRIDGE
The Pitt Building, Trumpington Street, Cambridge
CB2 1RP United Kingdom

CAMBRIDGE UNIVERSITY PRESS
40 West 20th Street, New York, NY 10011-4211,
USA http://www.cup.org

First published in Great Britain in 1998 by
Mitchell Beazley, an imprint of Reed Consumer
Books Limited, Michelin House, 81 Fulham Road,
London SW3 6RB

First published in the United States in 1999.

Printed in China by Toppan Printing Company

Typeset in Goudy, Helvetica, Helvetica Neue,
Symbol and Trifont in QuarkXpress™ [SE]

Library of Congress Cataloguing in Publication data
available

ISBN 0 521 64239 6 hardback

Endpapers: star-forming regions
in the Eagle nebula
Page 1: a massive storm on Saturn
Page 2: star trails around the Keck 1
telescope on Mauna Kea, Hawaii
Page 3: Ursa Major
Pages 4–5: the Pleiades
Right: comet Hyakutake in the
spring of 1996

CONTENTS

The Hubble Deep Field is an image of the far reaches of the observable universe. Each speck, except the bright star lower middle, is a galaxy crowded with millions of stars similar to our Sun. It is a picture representing the frontiers of knowledge.

INTRODUCTION

The universe is a wondrous and fascinating place. This book makes it possible, from the comfort of your armchair, to journey through the cosmos to investigate all aspects of the universe, from how to find your way round the night sky to understanding the nature of a black hole.

Assuming no previous knowledge of astronomy, top scientists from around the world explain the latest research and theories in a readable and very accessible manner. With extensive use of diagrams and original artwork, the structure and layout of the book enables you, whatever your level of knowledge, to follow an intriguing, exciting voyage of discovery.

The first question answered is 'Where are we?' Where do we fit into the universe? We no longer believe we are at the centre, but it is surprising to discover just how unimportant our position really is.

Beginning in our own back yard, we take a look at our solar system: the planets and other bodies that orbit our nearest star, the Sun. Completely up-to-date, the book includes information from the latest spacecraft and probes to be sent out to study both the Earth and our nearby neighbours. Boxes of key facts allow for quick reference.

From the solar system we journey out to the stars and examine all the weird and wonderful objects that exist, including white dwarfs, red giants, supernovae and black holes. The secrets of all the twinkling points of light seen at night are revealed.

Continuing to travel outward, we take a look at immense collections of stars, dust and gas – the galaxies. From the study of galaxies we cross over to cosmology, which is the study of the universe as a whole. The book examines such fundamental questions such as 'How did the universe come into being?' and 'What will happen to us and everything else within the universe in the future?'.

From the theories behind all the objects in the universe, we move to observing them. Specially commissioned, clear star maps are included with instructions on how anyone from anywhere on Earth can find their way around the night sky. In a unique approach, one easily recognizable pattern of stars, the constellation of Orion, is used throughout the book as a stepping stone to the rest of the sky.

Extensive cross referencing, the Astronomical Terms section on pages 10–11 and the glossary enable the reader to discover the meaning of any scientific term with ease. More knowledgeable readers will not be frustrated by continual definitions, while the newcomer is introduced to basic astronomical ideas with the aid of easily understandable analogies.

Throughout the book, complex concepts are described in a unique user-friendly approach. In addition, this book contains feature articles on exciting and controversial topics written by top scientists in the field. These articles vary from the search for other intelligent life to the threat of rogue asteroids.

This book opens the door to an incredibly fascinating place: the universe.

HOW TO USE THIS BOOK

Each of the six chapters begins with an overview of the subject and explores the fundamental theories involved, for example the chapter on the solar system starts with its formation, stars opens with the lifecycle of a star and the cosmology section opens with the Big Bang.

Within each of the six main chapters the text and captions cover the basics in a clear, jargon-free way, and information on the history of discoveries, more complex theories, satellites and scientific missions is explored in more detail in box

features. Specially commissioned artworks and diagrams illustrate both basic information and more complicated ideas as clearly as possible.

The chapter-by-chapter colour-coding and cross references make checking basic facts and related information quick and simple.

Astronomical Terms on pages 10–11 introduces the basic ingredients of the universe, and explains the systems that astronomers use to describe and measure distances, angles and temperatures. There is also an extensive glossary on pages 184–186.

Introduction
The Solar System
Stars
Galaxies
Cosmology
Professional Astronomy
The Night Sky

The chapter openers introduce the basic concepts that are covered in more detail on the following pages, and explore the fundamental theories behind them.

This colour bar indicates the start of each new chapter and is echoed in a strip along the bottom of each page in that chapter.

Specially commissioned, clearly annotated artworks are used to show both basic data and more complex information in a simple, easy-to-understand way.

Extended captions explain in detail the fundamental theories introduced on these spreads.

Box features treat separately great theories, the history of discoveries and the people associated with them.

Unique, colour-coded cross references lead quickly to detailed explanations of basic ideas and related subjects.

Each planet has clear graphics to show details of its size, and orbit and rotation.

The planet pages give information on satellites and major missions to the planets in boxes for easy reference.

Chapter-by-chapter colour coding enables you to find your way round the book easily and quickly.

ASTRONOMICAL TERMS

ASTRONOMICAL INGREDIENTS

Asteroid A rocky body between under 1km and 1000km in diameter that orbits the Sun. Most lie in the asteroid belt between Mars and Jupiter, although others occur throughout the solar system.

Comet A small, icy body orbiting the Sun in a long, eccentric orbit. Most comet nuclei are a few kilometres across, but their tails can extend for as much as 10 million kilometres.

Galaxy A collection of stars, gas and dust held together by gravity. Galaxies are classified as spiral, elliptical or irregular, according to their shape. They usually appear in groups known as clusters or larger groups called superclusters.

Meteor A grain of rock or dust a few millimetres across travelling through space at high speed, seen as a streak of light when burning high in the Earth's atmosphere.

Meteorite A meteor, perhaps a few metres across, that does not burn up in the Earth's atmosphere and hits the ground.

Meteoroid The collective term for all meteoritic bodies.

Micrometeorite An extremely small particle, less than 0.1mm in diameter. They are too small to produce shooting stars.

Nebula A cloud of interstellar gas and dust. Nebulae are detectable as emission nebulae, which glow; reflection nebulae, which scatter starlight; and dark nebulae, which obscure light from more distant stars or nebulae.

Planet A body in orbit around the Sun or other star. Pluto, with a diameter of 2290km, is the smallest in our solar system.

Satellite A body in orbit around a larger, parent body. Natural satellites of planets are called moons.

Star A luminous ball of plasma that shines by generating energy in its core by nuclear reactions.

Universe Everything.

ATOMIC INGREDIENTS

Atom The smallest quantity of an element that can take part in a chemical reaction.

Electron A stable elementary particle present in all atoms, orbiting the nucleus, usually in numbers equal to the number of protons in the nucleus.

Element A substance that consists of atoms with the same number of protons in its nucleus. Oxygen is an element; carbon dioxide is not.

Ion An electrically charged atom or group of atoms formed by the loss or gain of one or more electrons.

Molecule The most basic unit of a chemical compound, consisting of two or more atoms that are held together by chemical bonds.

Neutrino A stable, neutral elementary particle that travels at half the speed of light.

Neutron A neutral particle found in most atomic nuclei.

Nucleus A sub-atomic particle consisting of protons and, in most cases, neutrons.

Photon A quantum of electromagnetic radiation.

Plasma A hot ionized material consisting of nuclei and electrons. It is is the material present in the Sun and most stars.

Proton A stable, positively charged elementary particle, found in atomic nuclei.

LINEAR MEASUREMENTS

Metric measurements are used throughout this book. (A kilometre is 1093.6yds or just under ⅝ of a mile.)

One aspect of astronomy that might seem daunting is the huge distances involved. Our nearest neighbour, the Moon, is 376,280km away on average, the same distance as nearly 30 times around the Earth's equator. The nearest planet, Mars, is 56,000,000km away at its closest approach, the equivalent to 4390 circuits of the Earth's equator. Because the distances involved in astronomy are hard to comprehend, astronomers have defined their own distance scales.

Within the solar system they use **Astronomical Units** (**AU**). An AU is the average distance from the Earth to the Sun: and is equal to 149,597,870km. So, rather than saying that Jupiter is 778,340,000km from the Sun, this distance is described as 5.2 AU.

Beyond the solar system, AU are unwieldy to use for huge distances. Even the nearest star to the Sun, Proxima Centauri, is 25,000,000,000,000km away, or about 3,000,000 AU: so astronomers use a larger scale. This scale is based on the distance that light travels in a year – 9,461,000,000km. So Proxima Centauri is about 4.3 **light years** away. For smaller distances **light seconds** and **light minutes** are used.

For greater distances, astronomers use **parsecs**. This is equivalent to 3.26 light years. A **kiloparsec** (**Kpc**) is 1000 parsecs and a **megaparsec** (**Mpc**) is one million parsecs.

Light second
The distance that light travels in a second – 299,812.59km.
Light minute
The distance that light travels in a minute – 17,998,775km.
Light year
The distance that light travels in a year – 9,461,000,000km.
Parsec 30,842,860,000km.

DISTANCES OF OBJECTS FROM THE SUN

Object	Light years
Nearest star (Proxima Centauri)	4.3
Sirius	8.6
Betelgeuse	190
Centre of Milky Way	32,000
Distance across Milky Way	80,000
M31 (Andromeda galaxy)	2.3 million
Virgo cluster of galaxies	50 million
Farthest galaxies known	11 billion

CELESTIAL COORDINATES

If you look up into the night sky, the stars appear fixed at the same distance onto the inside of a huge upside-down pudding basin. In reality, all the stars are at hugely different distances from us; the pudding-basin effect is an optical illusion. The pudding-basin idea is used, however, in the concept of the 'celestial sphere', which is used for the night sky's system of coordinates.

Celestial sphere An imaginary huge sphere that is surrounding the Earth.

Celestial equator The projection of the Earth's equator onto the celestial sphere.

Celestial poles The projection of the Earth's poles onto the celestial sphere.

Ecliptic The projection of the Earth's orbit onto the celestial sphere. It also marks the yearly path of the Sun across the sky against the background of the stars.

Right Ascension Celestial coordinate analogous to longitude on the Earth.

Declination The angular distance of a celestial body north or south of the celestial equator. It is analagous to latitude on the Earth.

In addition, astronomers use a second system of coordinates that describe the positions of objects relative to the position of the observer.

Altitude The angle of a body above the observer's horizon ranging from 0° at the horizon to 90° at the zenith.

Azimuth The angle of a body from the observer's north point, ranging from 0° at due north to 180° at due south.

Meridian An imaginary line on the sky that runs due north–south and passes through the observer's zenith.

Zenith The point on the celestial sphere directly overhead.

ANGULAR MEASUREMENTS

Astronomers use angular measurements to describe the apparent distance between objects in the sky (angular separation), that is, the distance between objects as they appear on the celestial sphere.

The sky is divided into 360 degrees (°). Each degree is divided into 60 minutes, and each minute divided into 60 seconds. On the sky these are called arc minutes and arc seconds.

Closer to the Earth, this angular measurement is used to describe the size of objects, such as the Sun, which is 0.5°, or 30 arc seconds across.

1° = 60 arc minutes
1 arc minute = 60 arc seconds

Item	degrees	arc minutes	arc seconds
Moon	0.5	30	1800
Betelgeuse–Rigel	c.20	c.1200	c.72,000
Across Orion's belt	c.3	c.180	c.10,800

SCIENTIFIC NOTATION

Even when they are described in light years, the distances between stars and and galaxies are so great that astronomers abbreviate them as shown below.

1 billionth (US)	=	0.000000001	$= 10^{-9}$
1 millionth	=	0.000001	$= 10^{-6}$
1 thousandth	=	0.001	$= 10^{-3}$
1 hundredth	=	0.01	$= 10^{-2}$
1 tenth	=	0.1	$= 10^{-1}$
1 hundred	=	100	$= 10^{2}$
1 thousand	=	1000	$= 10^{3}$
1 million	=	1,000,000	$= 10^{6}$
1 billion (US)	=	1,000,000,000	$= 10^{9}$

In this way, the mean distance between the Earth and the Sun (149,000,000 km) is described as 1.49×10^{8} km.

TEMPERATURE

Throughout the book, we use degrees Celsius (°C). Water freezes at 0°C and boils at 100°C at sea level on the Earth. The Kelvin scale uses the same units as °C, but starts at a different point – the coldest that anything can ever be is 0K or –273.15°C. This is called absolute zero.

0K = –273.15°C or –459.67°F
255.37K = –17.77°C or 0°F
273.15K = 0°C or 32°F

AVERAGE DISTANCES OF THE PLANETS FROM THE SUN

Planet	AU	Light	km	Years at 100km/h
Mercury	0.4	3 mins, 20 secs	59,910,000	68.3 years
Venus	0.72	6 mins, 1 sec	108,210,000	123.4 years
Earth	1	8 mins, 19 secs	149,600,000	170.7 years
Mars	1.52	12 mins, 40 secs	227,940,000	260 years
Jupiter	5.2	43 mins, 16 secs	778,340,000	887.9 years
Saturn	9.54	1hr, 19mins, 20 secs	1,427,010,000	1627.9 years
Uranus	19.18	2hrs, 39mins, 31 secs	2,869,600,000	3273.6 years
Neptune	30.06	4hrs, 9mins, 59 secs	4,496,700,000	5129.7 years
Pluto	39.44	5hrs, 27mins, 59 secs	5,900,000,000	6370.6 years

WHERE ARE WE?

IT WAS ONLY IN 1923 THAT SCIENTISTS REALIZED THAT OUR GALAXY IS ONE AMONG MILLIONS AND THE UNIVERSE IS MUCH BIGGER THAN WE THOUGHT ...

Since then, bigger and better telescopes have enabled astronomers to look farther and farther out across the universe to uncover its structure and to discover our place in it.

The universe contains everything that exists. It is believed to have sprung almost instantly into being at the Big Bang as many as 15 billion years ago. It has been expanding ever since.

At the moment of the Big Bang, the universe was unimaginably small and incredibly hot. As it expanded and cooled, before it was one second old, the strange forces and particles seething around in it formed more recognizable ones such as gravitation, electromagnetism, and subatomic particles such as quarks, protons and neutrons.

Progress from then on was slower: the first simple atoms were formed after about 300,000 years and the until-then cloudy universe began to clear (like steam thins as water condenses out of it). Because of the action of gravity, these atoms (mainly helium and hydrogen) grouped into vast swirling filaments of gas separated by immense voids. Slightly denser areas in these filaments began to attract more gas and formed swirling clouds that eventually formed into what we would recognize as galaxies between two and four billion years after the Big Bang.

The existence of superclusters and voids is a major proof of the Big Bang theory and during the last 30 years, work has continued on mapping the galaxies in our region of the universe. Researchers have found that there are many strings and sheets of galaxies that lie around vast voids. The biggest void between superclusters that has been found so far lies in the constellation of Boötes. It is 300 million light years across and is surrounded by 'walls' and strings of superclusters containing thousands of galaxies – like an immense soap-bubble.

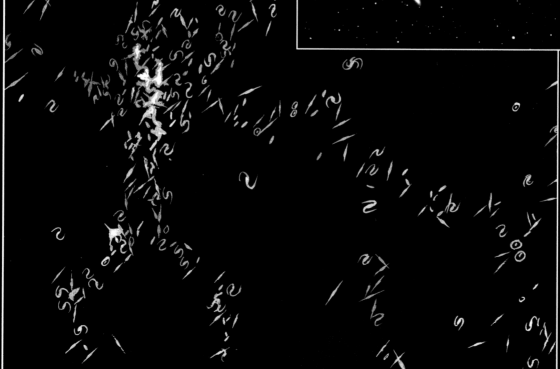

2▲ *The superclusters of galaxies themselves are made up of smaller groups of galaxies, although small is a relative term. Some are thought to contain as many as 1000 member-galaxies – our own local group contains between 30 and 40.*

1◄ *Most of the 'luminous' matter has condensed into superclusters of innumerable galaxies that seem to lie along and around filaments and in sheets that surround massive voids – like vast amounts of immensely long interconnected and tangled strands of macaroni.*

4◄ *Our solar system lies embedded within one of the spiral arms of the Milky Way, about two-thirds of the way between the nucleus and the edge of the galaxy. The faint white band of stars shows where the plane of the galaxy lies, although we cannot see all of the stars because there is too much dust and gas in between.*

3◄ *Our galaxy – the Milky Way – is a spiral, with a large, spherical nucleus from which the arms of stars, dust and gas radiate out and curve around. Seen edge-on, the galaxy would look like two fried eggs stuck back to back.*

It is estimated that there are 100 billion stars in our galaxy (the number of grains of rice it would take to fill a cathedral).

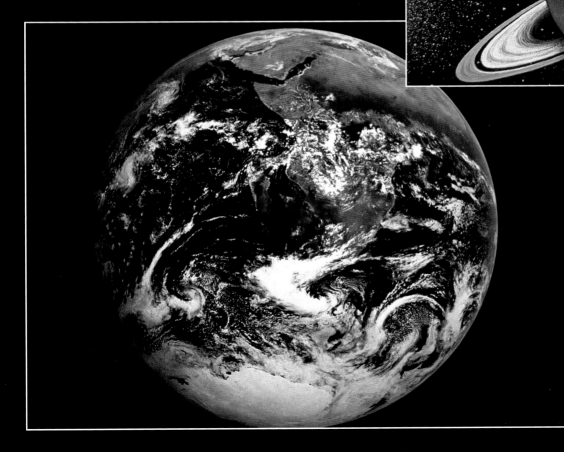

5▲ *The solar system consists of one star – the Sun – and nine planets with their satellites: the smaller, solid inner planets Mercury, Venus, the Earth and Mars; the gas giants Jupiter, Saturn, Uranus and Neptune; and tiny, rocky, frozen Pluto; as well as the minute asteroids, comets and meteors.*

Astronomers are now finding evidence that there are other planets around other stars.

6◄ *Our home, the Earth, is the third planet from the Sun. It takes 365 days and six hours to orbit the Sun at an average distance of nearly 150 million kilometres. It is the only planet in the solar system that can support life: those closer in to the Sun are too hot and those farther out are too cold.*

THE SOLAR SYSTEM

The Earth is one of nine planets orbiting a yellow dwarf star: the Sun. Our planet is rocky with large oceans of liquid water and an atmosphere of nitrogen, oxygen and carbon dioxide. It has one natural satellite: the Moon.

INNER SOLAR SYSTEM
This consists of the four terrestrial planets: Mercury, Venus, the Earth and Mars.

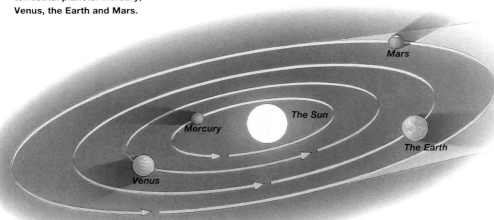

The other planets are diverse, inhospitable worlds, some with poisonous atmospheres, some with no solid surfaces, some baked to enormous temperatures and some cooled to well below freezing. There are planets with no moons, and planets with tens of moons, each moon a different world. Wandering among the planets are fragments of rock and ice: asteroids and comets left over from the formation of the solar system. Some are tens of kilometres across, some mere specks of dust. All these fascinating objects are members of the solar system, our astronomical back yard.

THE SOLAR SYSTEM
The comparative sizes of the Sun and the nine planets of our solar system.

OUTER SOLAR SYSTEM
The outer solar system consists of the four Jovian or gas giant planets (Jupiter, Saturn, Uranus and Neptune) plus Pluto. These planets orbit at the farthest distances from the Sun.

THE FORMATION

MANY THEORIES ABOUT THE ORIGIN OF THE SOLAR SYSTEM
HAVE BEEN PUT FORWARD, MODIFIED, DISCARDED, AND
SOMETIMES REVIVED, AS NEW EVIDENCE IS GATHERED.

Current scientific thinking is that the space between stars is not empty but filled with the interstellar medium: clouds of dust and gas. These clouds can be very tenuous, but in some regions there are denser clumps several light years wide. The birth of stars occurs when these enormous cold clouds collapse, and planets can then form around them.

THE BIRTH OF A STAR

It is generally believed that the Sun and the solar system formed at the same time, from an interstellar gas and dust cloud that astronomers call the solar nebula. The solar nebula itself formed when part of the interstellar medium was compressed due to a disturbance, probably the shock wave from a nearby supernova. It then went on to collapse under its own gravity, forming a turbulent disk of gas and dust.

As the cloud collapsed, the centre became very compressed, releasing heat. The thickness of the surrounding cloud meant that this heat was unable to escape and eventually temperatures rose so high that the dust near the centre vaporized, and the protostar that was to become our Sun formed. The initial collapse is thought to have taken place in less than 100,000 years. When the temperature and pressure became great enough, nuclear fusion of hydrogen into helium began and the Sun was born.

A SPINNING DUST CLOUD

All the planets in the solar system orbit the Sun in an anti-clockwise direction if viewed from above and the Sun itself rotates in this direction. In addition, most of the planets spin on their axes and the majority of planetary moons orbit their parent planet in this direction. This, coupled with the fact that the planets orbit the Sun almost in the same plane, strongly suggests the solar nebula was rotating.

If a collapsing dust and gas cloud is rotating, then centrifugal force will stop some gas and dust in the rotating disk from falling in to the centre and becoming part of the new star. The planets formed from this rotating disk.

THE DISK COOLS

The dust and gas within the disk were cooler the farther away they were from the protostar. The disk was also less dense and a great deal less compressed at the edges away from the centre.

As the material in the disk cooled, elements such as metal, rock and ice condensed out, just as water condenses out of steam from a kettle as it cools. These elements condensed out at different rates through the disk, depending on the distance from the hot proto-Sun. The metals condensed almost as soon as the disk formed; the rock condensed a few hundred million years later and the ice later still.

There would have been more icy material in the outer reaches of the newly forming solar system because of the lower temperatures. These outer reaches would also have a higher chance of retaining solid particles from the protosolar cloud because the temperatures never rose high enough to vaporize the original material. These particles probably included high-melting point material such as diamond and silicon carbide, grains of which astronomers now find in meteorites.

THE YOUNG PLANETS GROW

The condensed dust particles had a low-density, irregular structure that meant they tended to stick together when they collided with each other, forming larger particles. Eventually, particles the size of boulders or small asteroids were produced, which were massive enough to attract each other through gravity, so their growth rate accelerated.

The large objects tended to sweep up all of the solid matter in or close to their own orbit, so their final size was limited by the material available in their vicinity at their distance from the Sun. These bodies are known as planetesimals; those orbiting in the inner solar system were perhaps the size of the Moon and those in the outer solar system were up to 15 times the size of the Earth.

The reason for the size difference was that between the orbits of Mars and Jupiter the temperature fell to a level where ice could condense. It was easier for planets to form out of sticky, solid ice than liquid or vapour.

The rate at which collisions occurred was considerably less at large distances from the Sun because the disk of material was less dense, so the formation of these large planetesimals in the outer solar system took tens of millions of years rather than the few hundred thousand years required in the region of the Earth.

THE T TAURI STAGE

Theoretical models of the formation process show that the protosolar cloud must have been many times more massive than the contents of the present day solar system. At some stage, most of the mass must have been removed.

When they reach one million years old, stars similar to the Sun develop a very strong outflow of charged particles, like a massive version of today's solar wind. These stars are called T Tauri stars after the first one was observed in the constellation of Taurus.

◄ 14–15 The solar system

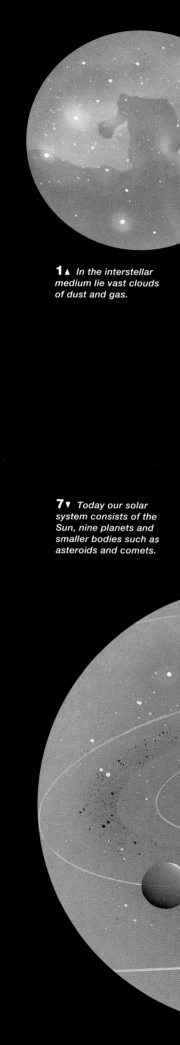

1▲ *In the interstellar medium lie vast clouds of dust and gas.*

7▼ *Today our solar system consists of the Sun, nine planets and smaller bodies such as asteroids and comets.*

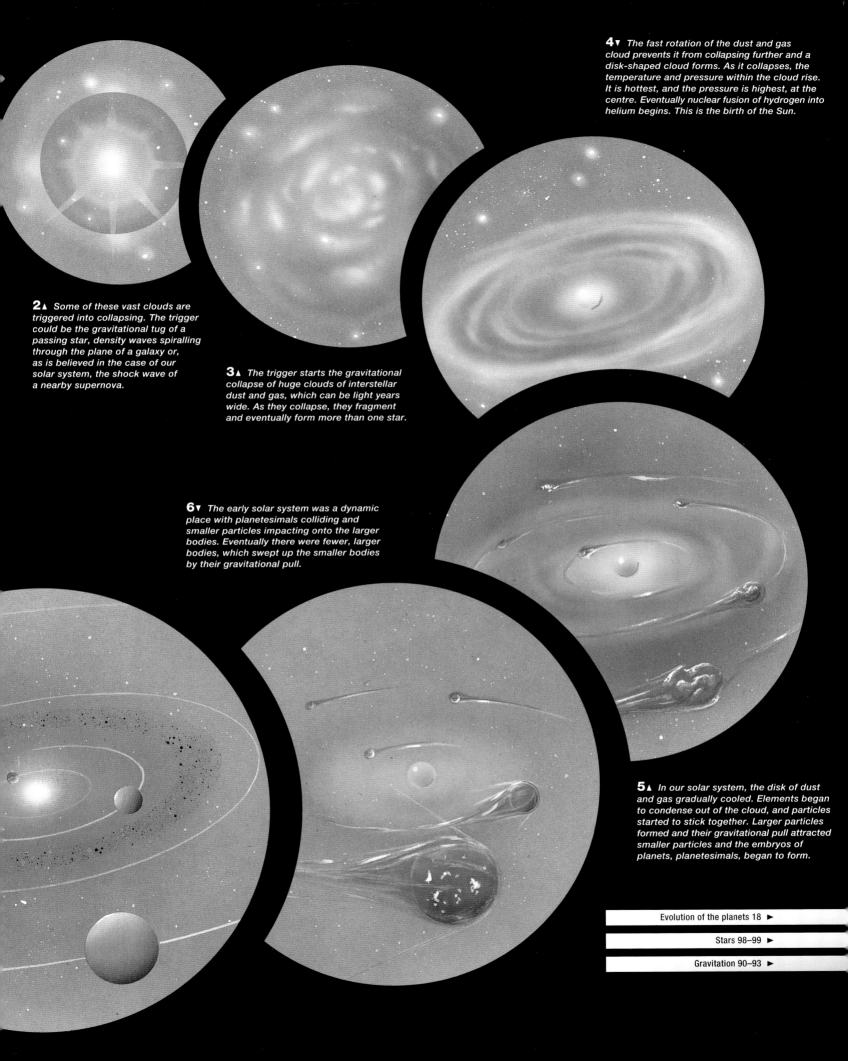

2▲ Some of these vast clouds are triggered into collapsing. The trigger could be the gravitational tug of a passing star, density waves spiralling through the plane of a galaxy or, as is believed in the case of our solar system, the shock wave of a nearby supernova.

3▲ The trigger starts the gravitational collapse of huge clouds of interstellar dust and gas, which can be light years wide. As they collapse, they fragment and eventually form more than one star.

4▼ The fast rotation of the dust and gas cloud prevents it from collapsing further and a disk-shaped cloud forms. As it collapses, the temperature and pressure within the cloud rise. It is hottest, and the pressure is highest, at the centre. Eventually nuclear fusion of hydrogen into helium begins. This is the birth of the Sun.

6▼ The early solar system was a dynamic place with planetesimals colliding and smaller particles impacting onto the larger bodies. Eventually there were fewer, larger bodies, which swept up the smaller bodies by their gravitational pull.

5▲ In our solar system, the disk of dust and gas gradually cooled. Elements began to condense out of the cloud, and particles started to stick together. Larger particles formed and their gravitational pull attracted smaller particles and the embryos of planets, planetesimals, began to form.

Evolution of the planets 18 ▶

Stars 98–99 ▶

Gravitation 90–93 ▶

During the Sun's T Tauri stage, all of the gas left in the Sun's protoplanetary nebula would have been swept away. The larger, more massive, outer planets would have been able to retain most of their atmospheres, but the smaller inner planets would have had their original atmosphere blown away.

EVOLUTION OF THE PLANETS

The planets and large satellites can be divided up into two distinct classes. Mercury, Venus, the Earth and Mars are the rocky terrestrial planets and Jupiter, Saturn, Uranus and Neptune are the large gaseous Jovian planets. Pluto can be classified together with the large icy satellites of the gas giants as a third type.

The terrestrial planets are closer to the Sun, have smaller masses and radii and are more dense than the Jovian planets. The Jovian planets are very big, low in density and have extensive satellite systems and rings.

As explained, the basic difference between the two families arose as a consequence of the temperature differences within the protosolar cloud. This allowed icy material to condense beyond the asteroid belt, producing massive, cold protoplanets, which efficiently collected large amounts of gas. The inner planets were too small and too hot to retain large amounts of original atmosphere after the strong winds of the Sun's T Tauri stage.

THE GAS GIANTS

The interesting question in the case of Jupiter and the other gas giants, Saturn, Uranus and Neptune, is the extent to which their atmospheres, and the planets as a whole, have the same composition as each other and as the Sun.

Recent evidence from measurements of methane by spectroscopy and direct sampling by the Galileo probe suggests that Jupiter's composition is rich in carbon in comparison to the Sun. This fits in well with the theory that the outer planets formed first from solid, ice-rich planetesimals and then, when sufficiently massive (about 10 times the mass of the Earth), proceeded to sweep up the gaseous part of the nebula.

It is possible, however, that the Jovian planets formed initially out of the gaseous nebula along with the Sun, almost like a multiple star system, and have been acquiring extra icy material ever since, in the form of a steady flux of comets.

THE TERRESTRIAL PLANETS

In general, the terrestrial planets differ from the gas giants of the outer solar system in that they were too hot to change volatile gases into liquids and solids and too small to retain the very abundant light elements hydrogen and helium when the T Tauri phase of the Sun removed the gaseous part of the nebula.

FORMATION OF PLANETS IN THE SOLAR SYSTEM

1 ▼ *In the original solar nebula, temperature and pressure are highest at the centre where the Sun eventually forms. Temperature and pressure fall as distance from the Sun increases and so the solar nebula is thicker farther out.*

Disk of dust and gas

Proto-Sun

Temperature and pressure decrease

Direction of rotation

2 ▼ *The temperature and pressure played a significant part in the final composition of the nine planets in our solar system.*

Proto-Sun

1AU: At the distance of the Earth the temperature rises to 2000K so all the original dust and gas are vaporized. As the solar nebula cools, water condenses into a liquid. Nearer the Sun, water remains as a vapour.

5AU: At the distance of Jupiter, the temperature is only 400K, so all the original dust and gas remain. As the solar nebula cools, water condenses into ice.

T Tauri stage of strong solar winds

Proto-Sun

1AU

The Earth

3◄ *The Sun went through what is known as the T Tauri stage when strong solar winds blew away small bodies and gas. The inner planets were not massive enough to hold on to their original atmospheres.*

4▼ *The planets farther away from the Sun took longer to form. So the outermost planets were smaller when the Sun went through its T Tauri stage. This explains why the* four gas giants get progressively smaller the farther away from the Sun they are because the solar wind removed the material from which the planets were forming.

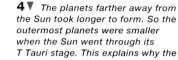

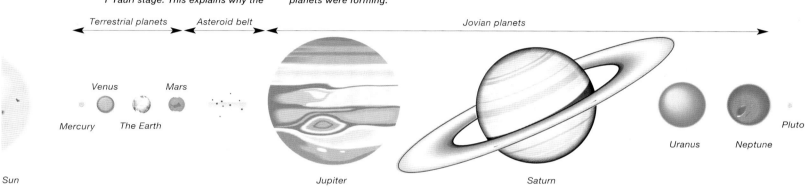

Terrestrial planets   Asteroid belt   Jovian planets

Venus   Mars

Mercury   The Earth

Sun

Jupiter   Saturn

Uranus   Neptune   Pluto

It is much more difficult, however, to explain the chemical differences between the four inner planets. This is not helped by the fact that our knowledge of the composition of these planets, even of the Earth, is far from complete.

FORMATION OF THE MOON

There have been various theories about the formation of the Moon, including the possibility that the Moon was 'spun off' the molten proto-Earth, or that the two bodies formed together as a 'double planet'.

However, the most favoured theory today is that the Moon formed as a planetesimal within the vicinity of the Earth, and then impacted onto the embryo Earth. The giant impact would have driven a huge amount of crustal material from the still-molten Earth into orbit where it would later have solidified and formed the Moon.

This theory also explains how the Moon comes to have little or no metallic core, since the Moon would have formed out of material mainly from the Earth's outer layers.

FORMATION OF ASTEROIDS

The asteroid belt occupies a part of the solar system between Mars and Jupiter. The largest asteroid is less than one per cent of the mass of the Moon, and all of them together are less than one-tenth of the Earth's mass. For a long time it was thought that the asteroids were the debris of a collision that destroyed a planet some time between its formation and the present day.

It now seems likely that no large planet ever formed between Mars and Jupiter, mainly because of the huge

pull of the latter's gravitational field. Most of the mass present in that region during the early days of the formation of the solar system was probably rotating in elliptical orbits and ended up colliding with the planets, their large satellites or the Sun.

FORMATION OF COMETS

For many years, all comets were believed to originate in the Oort cloud, a spherical region found in the far reaches of our solar system, approximately 100,000AU from the Sun. This may contain billions of icy planetesimals, which are perturbed out of orbit and into the inner solar system by the gravitational influence of nearby stars. However, scientists now think that this sort of perturbing mechanism does not fully explain the number of short period comets that are observed.

Recently, the idea of a belt of objects found inside the solar system, where stronger, more frequent perturbations from the planets themselves are possible, has come into favour. The existence of the Kuiper belt has been confirmed by the detection of several of its members, including one that is 150km in diameter, just beyond Neptune's orbit.

It is now estimated that this belt may contain up to one billion comets; their total mass is just one per cent of that of the Earth. Pluto and its moon Charon may be the largest members of the belt, rather than a proper planet-moon system. Similarly, Triton, Neptune's largest moon, which is orbiting backward and gradually spiralling inward around the planet, may also be a body like Pluto: a planet embryo captured into Neptune's orbit.

THE SUN

WITHOUT OUR NEAREST STAR, THE SUN, LIFE WOULD
NOT EXIST ON EARTH. IT IS THE ONLY STAR THAT IS
NEAR ENOUGH FOR US TO STUDY IN DETAIL.

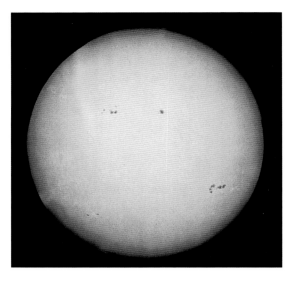

*The Sun, our nearest star, can be observed in ordinary
visible light by projecting the image onto white card. Often
the image will show dark areas that are known as sunspots.*

Like most stars, the Sun is made up mainly of hydrogen.
It is not a large star, in fact it is classified as a yellow
dwarf. However, compared to the Earth, the Sun is huge.

Different processes occur in different regions: the core
where energy is produced; the radiative and convective
zones; the photosphere; the chromosphere; and the corona.

THE POWER SOURCE

Within the core of the Sun the temperature and pressure are
so great that nuclear fusion of hydrogen into helium occurs.
This is known as hydrogen burning. The energy produced
by hydrogen burning is in the form of extremely high energy
photons that are at gamma ray wavelengths. This energy is
transported through the body of the Sun, cooling as it
moves from the centre, to become the light that reaches the
surface and escapes into space. During hydrogen burning,
protons change into neutrons, releasing photons and
particles called neutrinos. Neutrinos have no electric charge
and only a minute mass, if any.

Neutrinos do not interact easily with matter: they stream
out, unimpeded, from the Sun at the rate of about 10^{38} per
second. In fact, more than 150 million neutrinos pass
through every square centimetre of your body every second.

Astronomers have attempted to detect solar neutrinos,
but so far their experiments have detected only one-third of
the number that has been predicted.

THE RADIATIVE ZONE

Energy emitted at the core is transferred outward firstly by a
process called radiative diffusion. Because the density of
matter is so high at the centre of the Sun, an emitted photon
will not travel far before it hits another particle and is
reabsorbed. It can be re-emitted in any direction, but

because the temperature and pressure are so much higher
at the centre, photons will gradually diffuse outward.

If photons were unimpeded, they would reach the
photosphere in only two seconds, but instead it can take a
million years. Photons lose energy with each interaction: as
they move out, the pressure and temperature around them
gradually drop.

THE CONVECTIVE ZONE

In the outer layers of the Sun, convection is the main
process of transferring heat outward: material heated in the
lower layers rises to the top. As the material cools,
it circulates back down and is reheated. The tops of
the convection cells can be seen in ordinary light as
granulation – bright hot areas with cool dark edges.

THE PHOTOSPHERE

The Sun, in ordinary light, appears to have a clear edge.
This is because only a 500km layer of its atmosphere is at
the correct temperature to emit light at visible wavelengths:
a very small layer in comparison to its diameter.

THE CHROMOSPHERE

Above the photosphere the temperature rises steeply in a
region called the chromosphere. Astronomers can see it
during solar eclipses, or with spectrohelioscopes. This layer,
a few thousand kilometres thick, appears red because the
hydrogen atoms it contains are 'excited' by the high
temperatures and emit radiation at the red end of the visible
spectrum. A network of 'supergranules' exists here, similar
to granulation, but here the cells average 30,000km across
and correspond to circulation patterns that extend deep
into the Sun.

Bright jets of plasma called spicules rise to 10,000km
above the chromosphere and last for 5–10 minutes before
falling back or fading away. They form along the boundaries
of supergranules and are thought to be controlled by the
magnetic fields in these regions. They may also play a role
in heating the Sun's upper atmosphere.

NUCLEAR FUSION

**Four hydrogen atoms are changed
into one helium atom, producing a
small amount of energy. The Sun's
total power output, its luminosity,
is 3.9×10^{26} watts (equivalent to a
million, million, million, million
100-watt light bulbs). The amount
of energy that is produced in each
individual reaction is tiny, but the
Sun is converting 600,000 million
kilograms of hydrogen into helium
every second.**

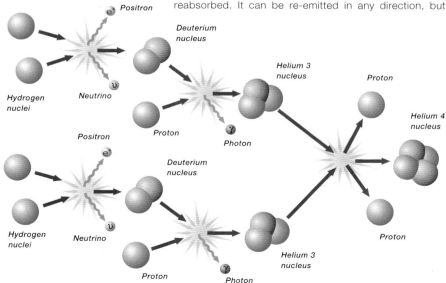

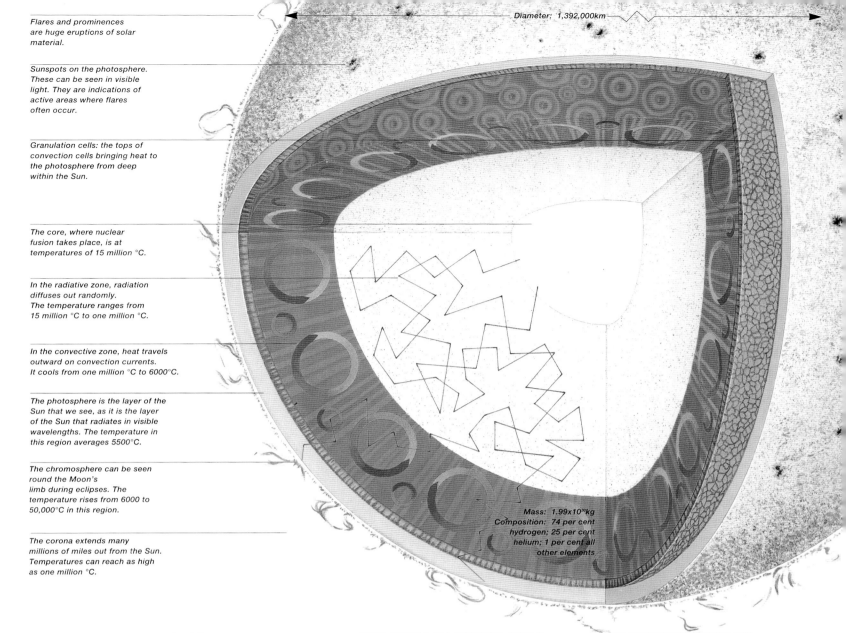

Flares and prominences are huge eruptions of solar material.

Diameter: 1,392,000km

Sunspots on the photosphere. These can be seen in visible light. They are indications of active areas where flares often occur.

Granulation cells: the tops of convection cells bringing heat to the photosphere from deep within the Sun.

The core, where nuclear fusion takes place, is at temperatures of 15 million °C.

In the radiative zone, radiation diffuses out randomly. The temperature ranges from 15 million °C to one million °C.

In the convective zone, heat travels outward on convection currents. It cools from one million °C to 6000°C.

The photosphere is the layer of the Sun that we see, as it is the layer of the Sun that radiates in visible wavelengths. The temperature in this region averages 5500°C.

The chromosphere can be seen round the Moon's limb during eclipses. The temperature rises from 6000 to 50,000°C in this region.

The corona extends many millions of miles out from the Sun. Temperatures can reach as high as one million °C.

Mass: 1.99x10²⁰kg
Composition: 74 per cent hydrogen; 25 per cent helium; 1 per cent all other elements

THE CORONA

The outer layer of the Sun, the corona, can be seen shining majestically during a total solar eclipse. It appears very white and shines mainly by scattered sunlight.

The corona extends for many millions of miles out from the Sun and during eclipses it can be seen that it has detailed structures in the form of rays, arches, plumes and streamers. Both the shape and nature of these features change with the sunspot cycle. The corona is both very tenuous and very hot. The temperature rises through the chromosphere, through a narrow band called the transition region, into the corona where it can reach one million °C. The cause of this rise in temperature is not yet completely understood but astronomers think that the action of the Sun's magnetic field must be involved in some way.

Less dense regions of the corona, which astronomers call coronal holes, appear where the Sun's magnetic field opens out to interplanetary space rather than looping back down. These areas are believed to be the major source of the solar wind, where energetic charged particles, mainly protons and electrons, stream outward into the interplanetary medium.

THE SUN'S MAGNETIC FIELD

Overall, the Sun's magnetic field is roughly the same strength as the Earth's, but the mechanism is entirely different. The Sun is not a solid body, but a plasma created by the heat of the Sun removing the electrons of hydrogen atoms to leave negatively charged electrons and positively charged ions.

Magnetic fields can be created by the motion of electrically charged particles and the Sun's turbulence and rotation create localized magnetic fields. As the Sun rotates, the magnetic field lines get 'trapped' and move around with the

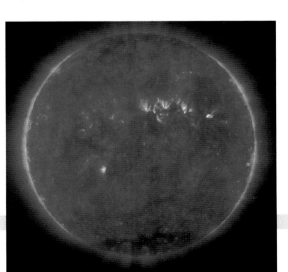

THE STRUCTURE OF THE SUN

The Sun is a lower main sequence star in which energy is fuelled by nuclear fusion in the core and brought out to the surface firstly by a radiative process, then by convection. These processes are not sharply defined but overlap.

The outer corona in ultraviolet. The bright regions are areas of intense magnetism. This part of the corona is at a temperature of one million °C.

Sun's rotation. As the top layers bubble with convection, the magnetic field lines get twisted up. As they get squashed together, the magnetic field strength increases in these areas, and these intense pockets of magnetic field cause many of the phenomena seen on the Sun.

SUNSPOTS

If you project an image of the Sun onto a piece of white paper, you may see dark spots on the Sun's disk. These are sunspots, first seen by Galileo and his contemporaries in the 17th century, although large naked-eye spots had been recorded by Chinese astronomers hundreds of years before.

Larger sunspots have a darker, umbra, region, with a lighter, penumbra, surround. Some spots – pores – are tiny while others can be hundreds of thousands of kilometres across and are visible to the naked eye. Most spots appear in pairs, but often very complex groups emerge.

Galileo used the movement of the spots across the solar disk to work out that the Sun rotates in about four weeks. Later observers found that different parts of the Sun rotate at different rates (as it is not a solid body): the equatorial regions rotate about once every 25 days; at the poles, it can take about 35 days to rotate.

BUNDLES OF MAGNETIC FIELD

Sunspots occur where magnetic field lines break out of and loop back into the photosphere. We see them because they are cooler areas of the photosphere and appear darker than surrounding regions. They can be more than 1000°C cooler, although they are still extremely hot at about 4800°C. Because they are regions of very intense magnetic fields, and the electrically charged plasma cannot easily cross magnetic field lines, the flow of plasma and heat is restricted. Some sunspots have magnetic fields as high as 4000 Gauss, compared to the solar average of 1 Gauss.

THE SUNSPOT CYCLE

The number of sunspots varies over roughly an 11-year period known as the sunspot cycle. In fact the sunspot cycle is part of a longer 22-year solar cycle. Many of the other phenomena observed vary with the sunspot cycle.

A pair of sunspots will have different magnetic polarity, and usually all the leading spots in one hemisphere will

Some sunspot groups can cover very large areas of the photosphere and can become beautifully intricate with dark, cooler umbra areas merging with the lighter, less cool outer penumbra regions.

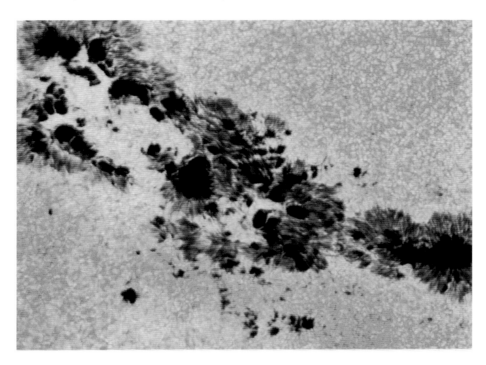

Right: Observing the Sun with a magnetogram shows up regions of different magnetic polarity. In this image the areas of intense magnetic fields within the sunspots can clearly be seen. Yellow regions are north poles and blue and pink regions are south poles.

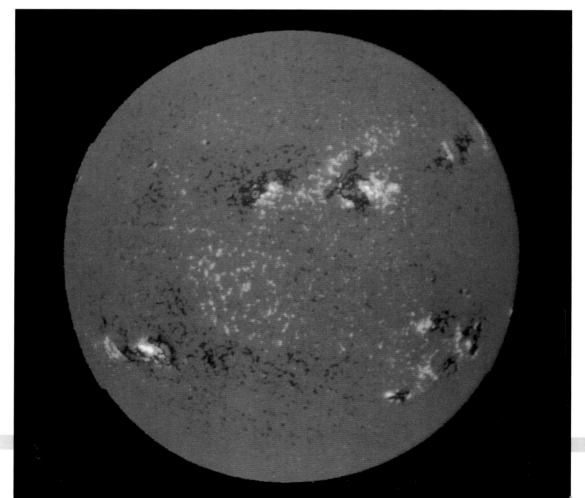

have the same magnetic polarity. About one year after solar maximum (the peak time of sunspot activity) the magnetic polarity changes. This magnetic solar cycle takes twice the sunspot cycle to complete.

OTHER PHENOMENA

When looking at the Sun across the whole electromagnetic spectrum, astronomers can see a host of phenomena. Prominences, huge arches of solar material, can be seen at the limb of the Sun. When viewed face-on against the bright photosphere, they are known as filaments.

Spicules, narrow jets of gas, can also be observed at the limb of the Sun. They move at around 20–30km a second from the lower chromosphere into the inner corona, and fall back or fade away after about 5–10 minutes.

High above the sunspot areas are faculae and plages: bright patches of hotter gas. Faculae occur in the upper photosphere while plages appear in the chromosphere. Faculae can be seen in white light as bright patches at the solar limb. They occur across the entire disk, but because the central disk is so bright, it masks them.

The Sun's limb appears darker in visible wavelengths because you look through a greater thickness of the cooler photosphere. At X-ray and radio wavelengths, however, the limb appears brighter because these types of radiation are emitted from the hotter chromosphere and inner corona.

Solar flares, intense outpourings of solar energy, usually occur in complex sunspot groups. Shock waves from flares spread out through the corona, sometimes causing large amounts of plasma to be ejected. The material that has been ejected joins the solar wind and can cause enhanced auroral activity and magnetic storms on the Earth.

THE ENIGMATIC SUN

Despite a great deal of research, the Sun still keeps some secrets. The cause of sunspots is not entirely understood, and no theory has been able to explain their disappearance between 1645 and 1715.

It is obvious that the Sun has a tremendous effect on the Earth, but the subtle changes that occur due to the solar cycle are still being investigated with a range of instruments and spacecraft that cover the electromagnetic spectrum. Even the vibrating solar interior is being probed by solar seismology in an attempt to unravel the Sun's mysteries.

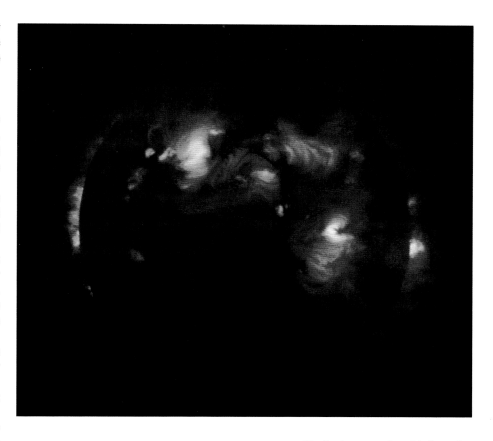

The Sun's corona viewed in X-rays by the Japanese satellite Yohkoh. The eruptions of flares and prominences occur over 'active areas' where sunspots often lie. The darker regions of the image are at about 5000°C and the brightest at about one million °C.

MISSIONS TO THE SUN

Name	Date	Aims and achievements
Yohkoh	1991	Monitors high-temperature phenomena taking place in the Sun's outer layers by using the X-ray part of the electromagnetic spectrum.
Ulysses	1994	Measures solar X-rays, radio waves and plasma waves. Flew under south pole and measured the magnetic field.
SOHO	1995	The Solar and Heliospheric Observatory studies the Sun's interior and also the solar wind.

THE SUNSPOT CYCLE

Sunspots rarely appear above more than 40° either side of the solar equator. The reason for this is not yet fully understood. E W Maunder (1851–1928) noticed that sunspots appear at high latitudes at the beginning of a solar cycle, but appear at lower and lower latitudes as the cycle progresses. This is thought to be a result of the 'winding up' of the Sun's magnetic field as the solar cycle progresses. Maunder's diagram showing this effect is known as the Maunder butterfly diagram. Spots of the new cycle can appear at high latitudes before the old cycle has finished.

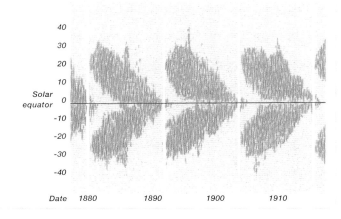

MERCURY

MERCURY IS THE FASTEST PLANET IN THE SOLAR
SYSTEM AND TAKES JUST 88 DAYS TO ORBIT ITS
MASSIVE NEIGHBOUR, THE SUN.

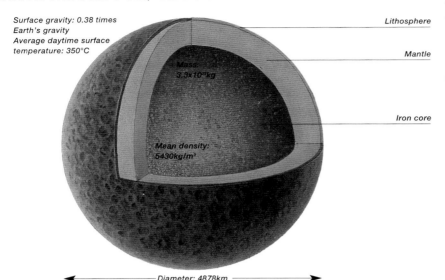

Surface gravity: 0.38 times
Earth's gravity
Average daytime surface
temperature: 350°C

Mass:
3.3x10²³kg

Mean density:
5430kg/m³

Lithosphere

Mantle

Iron core

◄——————— Diameter: 4878km ———————►

VITAL STATISTICS

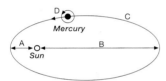

D
C
Mercury
A
B
Sun

A: 46,000,000km
B: 69,800,000km
C: orbits in 87.97 Earth days
D: rotates in 58.65 Earth days

Earth   Mercury

*Mercury's diameter is 0.38 times
the size of the Earth's*

ecliptic
orbit   equator

Angle of equator to ecliptic: 0°
Angle of orbit to ecliptic: 7°

Satellites: none

◄ 18–19 Formation of the planets

◄ 21 The corona and solar wind

Mercury is the closest planet to the Sun in our solar system and suffers the widest extremes of temperature. At Mercurian noon, when the Sun is directly overhead, the temperature can soar to as high as 470°C while during the long Mercurian night it can plunge to well below −173°C.

MOTION AND ROTATION

Although Mercury can appear to be brighter than the brightest star, and its existence has been known since the dawn of history, it is notoriously difficult to observe. This is because it is always so close to the Sun in the sky. The angle between Mercury and the Sun can never exceed 28°. This means that Mercury is lost in the Sun's glare because it sets no more than two hours after the Sun and rises no more than two hours before it. Once or twice a year, you may be able to see Mercury shining like a bright star close to the western horizon after sunset or close to the eastern horizon before sunrise.

Mercury orbits the Sun in only 88 Earth days and undertakes the Earth at intervals of, on average, 115.88 days. On these occasions, Mercury lies between the Sun and the Earth, but because of the tilt of its orbit, usually passes above or below the Sun when viewed from the Earth. Occasionally, when the alignment is right, Mercury passes directly in front of the Sun and can be seen as a small dot moving slowly across the face of the Sun: such an event is called a transit. The alignments that allow transits of Mercury to take place occur only in the months of May or November. Dates of recent and forthcoming transits are: 6 November 1993; 15 November 1999; 7 May 2003; 8 November 2006; 9 May 2016; and 11 November 2019.

Until the 1960s, most astronomers believed that Mercury took exactly the same time to rotate on its axis as it took to orbit the Sun: one hemisphere would always face toward the Sun and constantly suffer its boiling heat, while the other was in constant darkness. However, radar measurements carried out since then have shown that this is not the case. Mercury rotates every 58.65 days, which is precisely two-thirds of its orbital period or year.

This rate is still slow in relation to the orbital time and means that from Mercury, the Sun appears to move slowly across the Mercurian sky and one 'day' lasts as long as 176 Earth days, or two Mercurian years.

SURFACE FEATURES

When Mercury was newly formed, heat was generated by a number of factors: massive impacts on its surface by meteorites, asteroids and comets, the tidal effects of the Sun's gravity and the decay of radioactive elements. The heat melted the planet's interior and allowed heavy elements, such as iron, to sink toward its centre. Then, as Mercury cooled, it contracted slightly. This caused its surface to become wrinkled, so producing features such as the winding 'lobate scarps': curving cliff-like features with heights of up to 4km, which stretch for hundreds of kilometres across the planet's surface.

Images returned by the Mariner 10 (1974) spacecraft revealed that most of Mercury's surface is heavily cratered

The largest impact basin on Mercury is the Caloris Basin (half of which is shown here). It is 1300km in diameter and the surrounding area shows the shock waves generated by the impact. This image is a mosaic created by images taken from Mariner 10 in 1974.

and many of the craters are more than 200km wide. The largest feature seen by Mariner 10 was the Caloris Basin, an impact basin measuring 1300km across, which was surrounded by concentric ridges up to 2km high. Earth-based radar images have revealed three more large impact basins.

As on the Moon, some of the craters are surrounded by lighter-coloured ejecta – material splashed out by the impacts that formed these features.

In contrast, there are also some gently undulating plains where there are fewer craters. These are thought to have been produced by lava flows and are relatively crater-free because they took place after the period when most impacts occurred.

INTERIOR AND MAGNETIC FIELD

Mercury's mean density (5.43 times that of water) is similar to that of the Earth. For such a small planet to have so high a density, it must contain a lot of iron and astronomers believe that Mercury has twice as much iron, by proportion, as any other planet. Its iron core, which is thought to extend out to three-quarters of its entire radius, is surrounded by a mantle of rock and a thick crust.

The strength of the magnetic field at Mercury's surface is very low: only about one per cent that of the Earth's. This is only just strong enough to deflect most of the incoming solar wind and to form a magnetosphere around the planet.

Mariner 10's discovery of the magnetic field came as a surprise to most astronomers. According to conventional theory, a planet can only sustain a magnetic field if it has an electrically conductive liquid interior and rotates rapidly on its axis. Although Mercury has a large iron core, this ought to have cooled and solidified by now because of the planet's small size. The presence of a magnetic field suggests that at least part of the deep interior must still be liquid. Even if this is the case, Mercury's slow rotation makes the presence of a magnetic field rather puzzling.

MERCURY'S ATMOSPHERE

Mercury has a very tenuous and thin atmosphere with a ground pressure of about 2×10^{-9} millibars – only two-trillionths of the atmospheric pressure at the surface of the Earth. This means that if a human being were to stand unprotected on the surface of Mercury, the air in his or her lungs would expand and escape.

The atmosphere consists mainly of atoms and ions of hydrogen and helium, captured temporarily from

the solar wind. It also contains minute quantities of atomic oxygen, sodium and potassium. As these atoms and ions escape back into space, they are replenished by the supply of the incoming solar wind.

A mosaic of Mercury created from images taken by the Mariner 10 spacecraft on its outgoing journey in 1974.

SOLVING MERCURY'S RIDDLES

The existence of Mercury has been known for many centuries and it was the Romans who appropriately named the planet after the fleet-footed messenger of the gods. However, little more was learned about the planet until the mid-19th century when the French astronomer Urbain Leverrier realized that as Mercury moves rapidly along its elongated orbit round the Sun, the orbit itself slowly rotates (precesses) round the Sun. While most of Mercury's precession is caused by the gravitational pulls of the other planets and could be explained by Newton's theory of gravity, a small excess precession could not be accounted for. Leverrier proposed that the excess motion might be caused by the pull of an unseen planet, which he called Vulcan, closer to the Sun than Mercury. It is now known for certain that Vulcan does not exist.

A complete explanation for the behaviour of Mercury's orbit was provided by Einstein's General Theory of Relativity, which was published in 1915. For all of the other planets, Einstein's theory gives almost exactly the same results as Newton's theory but, because Mercury passes closer to the Sun than any other planet, the effects of relativity cause just the right amount of orbital motion to account for the excess movement of Mercury's orbit.

VENUS

VENUS, OUR SISTER PLANET, ONCE CONSIDERED
A TROPICAL PARADISE, IS NOW KNOWN TO BE A
BURNING, SUFFOCATING, DEADLY PLACE.

Surface gravity: 0.903 times Earth's gravity
Average surface temperature: 480°C

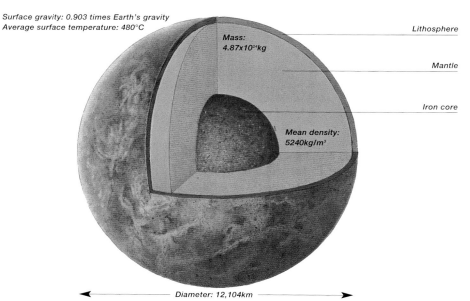

Mass:
4.87x10²⁴kg

Lithosphere

Mantle

Iron core

Mean density:
5240kg/m³

◄——— *Diameter: 12,104km* ———►

VITAL STATISTICS

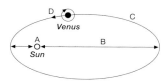

A: 107,500,000km
B: 108,900,000km
C: orbits in 224.7 Earth days
D: rotates in 243.01 Earth days

Earth   *Venus*

Venus' diameter is 0.95 times the
size of the Earth's

ecliptic
orbit   *equator*

Angle of equator to ecliptic: 177.4°
Angle of orbit to ecliptic: 3.4°

Satellites: none

◄ 18–19 Formation of the planets

Veiled as it is beneath its clouds, Venus remained a mystery until very recently. The planet that appears in our sky as the evening and morning star has inspired generations of romantics over the centuries.

A LIFELESS WORLD

An analysis of the sunlight reflected from the Venusian clouds revealed that the atmosphere was chiefly composed of carbon dioxide, and radio measurements suggested that the surface was extremely hot. The true face of Venus was finally beginning to emerge. Space probe soundings of the atmosphere and surface revealed a world completely devoid of all forms of water and confirmed searing surface temperatures that reached as high as 480°C.

As if this were not enough, the dense atmosphere crushes down on the planet with a pressure 90 times that at the Earth's surface, and the clouds are made of sulphuric acid droplets. Indeed, a human being standing unprotected on the surface would be simultaneously roasted, crushed and asphyxiated. So much for the Goddess of Beauty. Of all the planets in the solar system, Venus is the one that most closely approaches the classical picture of Hell.

VENUS AND EARTH: DISSIMILAR SISTERS

Until recently, Venus and the Earth were considered to be sister planets. They are a similar size, and were formed in much the same place and out of the same elements in space. Why then are they so very different?

Liquid water is, of course, the essential ingredient for life as we know it. So without any water source, it is extremely unlikely that any form of life ever existed on the planet.

Venus' proximity to the Sun means that it probably started out with less water than the Earth did; moreover any water there would probably have existed as water vapour rather than as liquid.

Even after the planets formed, the more intense solar radiation on Venus would have driven what little water remained from its atmosphere by breaking up the water molecules into their constituent parts of hydrogen and oxygen. Hydrogen is a very light gas and would have escaped off into space, while the oxygen would have been absorbed by the planet's surface. Even the rain of watery, icy comets that must have impacted Venus during its history was apparently insufficient to prevent the planet drying out.

VENUS' ATMOSPHERE AND ACID RAIN

When the interiors of both Venus and Earth heated up from radioactivity, a great deal of volcanic activity occurred, causing vast amounts of carbon dioxide to be released. On the Earth, the oceans dissolved some of this gas, and carbonate rocks were formed, but on Venus, there were no oceans so the carbon dioxide stayed in the atmosphere.

Findings from the US Pioneer Venus 2 (1978) spacecraft, which parachuted through the atmosphere, established that Venus has sulphuric acid clouds concentrated in a layer at heights of 48–58km above the surface. Drops of the acid develop just like drops of water in our own clouds and when they are large enough, they fall as acid rain.

However, this corrosive rain never reaches Venus' surface because the temperature difference (13.3°C at the top of the clouds and an oven-like 220°C underneath them) causes them to evaporate at about 31km above the ground. Below this level, the Venera and Pioneer probes have revealed that the atmosphere is remarkably clear, although the surface lies under a permanent overcast.

MISSIONS TO VENUS

Name	Date	Aims and achievements
Mariner 2	1962	Fly-by confirming the hostile nature of Venus' surface.
Venera 7	1970	First controlled landing on another planet.
Mariner 10	1974	First close-up pictures.
Venera 9	1975	First pictures from the surface.
Pioneer Venus 1	1978	Orbiter: first radar, ultraviolet and infrared observations.
Pioneer Venus 2	1978	First probes to measure Venus' atmosphere.
Magellan	1990	Global radar mapping of all of Venus' surface, down to objects 120m across.

The northern hemisphere of the surface of Venus created from radar images mainly taken by the Magellan spacecraft in 1989. The north pole is in the centre and the bright feature below this is Maxwell Montes, the highest mountain, 11km high. Simulated colour is used.

The clouds rotate 60 times faster than the planet, taking only four days to go around Venus once, and this rapid motion is driven by the heating of the atmosphere by the Sun. Pictures of Venus taken in ultraviolet light reveal many shapes in the clouds, including a global V-shaped pattern, which results from the mixing of two large swirling currents in the rapidly rotating atmosphere.

SEASONS AND TEMPERATURE

As Venus is not tilted on its axis like the Earth or Mars, it does not experience changing seasons. Throughout the Venusian year the equatorial atmosphere receives much more heat from the Sun than polar regions. However, warm air is efficiently transported to the cooler polar regions by giant convection currents, which pass through the dense atmosphere, reducing the temperature differences between the equator and poles to as little as a few degrees. The hot surface temperatures are exacerbated by the greenhouse effect and can reach as high as 480°C.

INTERIOR AND MAGNETIC FIELD

The internal structure of Venus is probably much the same as the Earth's, with a nickel-iron core surrounded by a silicate mantle. Nevertheless, the Mariner 2 (1962) probe discovered that Venus has a much weaker magnetic field than the Earth, suggesting that it may not have a liquid outer core.

The lack of a strong magnetic field around Venus may also be a result of the planet's slow rotation of 243 Earth days. Venus also rotates backwards as compared to the Earth and the other planets (ie, clockwise as seen from the north) – the reason for this is a mystery.

MAPPING VENUS

Since we cannot see the surface of Venus from above the clouds, scientists have had to resort to mapping the planet by radar. Radio waves that are beamed from orbiting spacecraft like Pioneer, Venera and Magellan pass directly through the clouds, bounce off the planet's surface, and the resulting echoes can be processed into images.

Pioneer Venus produced the first global relief map of Venus and revealed that the landscape is characterized by extensive areas of upland rolling plains (65 per cent), less extensive lowlands (27 per cent) and even smaller areas of highlands rising between 2–11km above the average surface level.

The Earth's internal structure 31 ►

The Earth's magnetosphere 32 ►

The Earth's atmosphere 36 ►

Seasons and the Earth's orbit 37 ►

The electromagnetic spectrum 140–141 ►

Robotic missions 148–149 ►

Observing the planets 160–161 ►

A radar image of the volcano Maat Mons, taken by the Magellan space craft in 1990. Maat Mons, 5km high, lies on the eastern edge of a large highland area of Venus, Aphrodite Terra, in the southern hemisphere. (Vertical relief in the image has been exaggerated 10 times.)

Most striking is the highland region of Aphrodite Terra, named after the Greek goddess of beauty, which snakes its way around the equator. To the east of this area is the volcano of Maat Mons (above).

The northern hemisphere is dominated by Ishtar Terra, named after the Babylonian goddess of love, which contains the Maxwell Montes – the highest mountains on Venus, which rise 11km above the average surface level of the planet. These were named after the physicist James Clerk Maxwell. There are also highlands that are more isolated such as Beta Regio, which is now known to be largely volcanic in origin.

A VOLCANIC SCENE

The lowland plains are entirely volcanic in origin and are made up of thousands of 'small' volcanoes and vast solidified lava flows. These volcanoes usually measure a few kilometres across, and often occur in clusters. The lava flows are more difficult to discern individually, but from images, many of the youngest examples can be picked out as they produce strong radar echoes and appear bright in the images. These flows cover large areas of the planet, stretching from tens to hundreds of square kilometres.

There are also many larger volcanoes on Venus, which can measure up to a few hundred kilometres across. These large volcanoes often sit astride major networks of faults and are characterized by hundreds of individual lava flows that lie radially around central vents, like spokes on a wheel.

Much of Venus's lava is likely to be of the fluid basaltic (dark volcanic) type like that of the Earth, Moon and Mars. Indeed, the surface landings of the Venera and Vega series have measured basaltic rock compositions at several locations. In addition, pictures returned by the Venera 9, 10 (1975), 13 and 14 (1982) landers show rocks that resemble the surfaces of lava flows that have been found on the Earth.

CORONAE

Unique to Venus are the coronae: curious circular 'blisters' that formed as plumes of molten rock rose to the surface. The coronae were seen for the first time in Venera 15 and 16 (1983) radar images. They are distinguished by concentric patterns of ridges and grooves that surround the volcanoes and lava flows often found at the centre of these features.

Despite the plethora of volcanic features on Venus, scientists do not know if any volcanoes are active today. Rather than having a global system of volcanoes strung along plate boundaries like the Earth, Venus has no plates and is dominated by 'hot-spots' like the coronae. The absence of plate tectonics on Venus may be due to the lack of water, which on Earth lowers the melting temperature of rocks and acts indirectly in the 'lubrication' of plate motion.

FAULTS AND FRACTURES

Venus is criss-crossed by thousands of faults and fractures, many of which run for thousands of kilometres

◄ 18–19 Formation of the planets

GREENHOUSE EFFECT

The air within a closed greenhouse on a summer's day is always warmer than the air outside. This is because the Sun's energy (or radiation) passes through the greenhouse glass, heats up the objects inside which radiate the energy back, but at longer wavelengths than the Sun's original energy. The longer wavelengths of radiation cannot get back through the glass, so the temperature within the greenhouse rises.

Some 'greenhouse gases' can act like the panes of glass, trapping heat at the surfaces of planets. Venus has a lot of carbon dioxide, an efficient greenhouse gas, in its atmosphere causing a very ferocious greenhouse effect on Venus.

The effect of increasing levels of carbon dioxide and other greenhouse gases in the Earth's atmosphere on global temperatures (global warming) has given cause for concern in recent decades. The study of Venus provides useful information on this process.

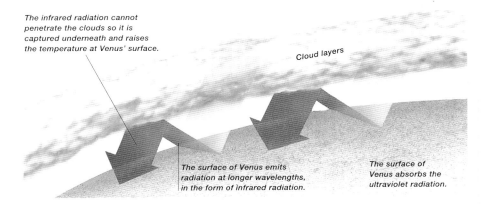

Ultraviolet energy from the Sun penetrates the thick cloud layers of Venus.

The infrared radiation cannot penetrate the clouds so it is captured underneath and raises the temperature at Venus' surface.

Cloud layers

The surface of Venus emits radiation at longer wavelengths, in the form of infrared radiation.

The surface of Venus absorbs the ultraviolet radiation.

---

across the surface. Some of the most recent and concentrated fracturing is found in Aphrodite Terra and Beta Regio where limited crustal extension and rifting has taken place.

There are also networks of ridge belts on the planet where the crust of Venus has been wrinkled up like a carpet by compression. Even greater compression has resulted in the mountain belts that surround Ishtar Terra in the north.

The complexity of ridges and fractures is greatest in the intricate ridged terrain, or tesserae, which have a long history of folding and faulting. Tesserae are very common in the area known as AphroditeTerra and around Ishtar Terra, and they also occur in isolated patches such as Alpha Regio. Regions of tesserae were flooded and closed in by the lava flows from the plains, and represent the oldest areas of exposed crust found on the surface of the planet.

IMPACT CRATERS

Over 800 impact craters have been counted in the Magellan images, and most appear to be remarkably fresh. They are surrounded by bright ejecta blankets spat out from the crater by the force of the impact. It is believed that they may have given way to the dark comet-like haloes, which may stretch for hundreds of kilometres across the surface in a westerly direction with the crater at their apex. These haloes were probably formed by surface dust blown high into the atmosphere by the impacts and then carried west by the atmospheric circulation before settling out.

All these craters seem to be on top of the plains and any older craters have been completely erased by the volcanic activity that formed the plains. Many scientists believe that the plains were formed in a global resurfacing event several hundred million years ago, erasing any older craters. However, others prefer the theory of continual resurfacing, and Venus's geological history remains controversial.

Venus can be seen shining like a bright star low in the west after sunset or in the east before sunrise. This photograph shows Venus to the right of a full moon with the constellation of Orion on the left.

THE EARTH

EARTH, AT 4.6 BILLION YEARS OLD, IS THE ONLY PLANET
IN OUR SOLAR SYSTEM KNOWN TO SUSTAIN LIFE. IT
ORBITS THE SUN AT NEARLY 2000KM PER HOUR.

VITAL STATISTICS

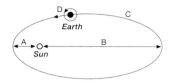

A: 147,100,000km
B: 151,100,000km
C: orbits in 365.26 Earth days
D: rotates in 23hr 56min 4.1sec

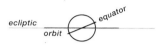

Angle of equator to ecliptic: 23.5°
Angle of orbit to ecliptic: 0°

Satellites: one

The Earth is a unique planet in our solar system. Its complex processes at work beneath the surface and its abundant supply of liquid water in the form of oceans provide the source for life. Earth sustains a perfect balance of temperature, atmosphere and pressure without which life could not exist.

THE EARTH AS A PLANET

Most ancient cultures understandably regarded the Earth as the centre of creation. Although a few Greek philosophers such as Aristarchus (c. 310–230 BC) claimed that the Earth orbits the Sun, most believed the Earth to be immobile, with the cosmos wheeling around it. Only after the work of the astronomer Nicolaus Copernicus in 1543, did people accept that the Earth and the other planets move round the Sun.

Similarly, it took time for people to believe that the Earth is round. Simple observations, such as ships passing out of sight below the horizon, showed that our planet is curved. Greek and Egyptians were first to make such claims but it was only after the globe had been circumnavigated in the 16th century that people accepted that Earth is truly a sphere.

At the end of the 18th century, Earth's age was called into question. Geologist James Hutton noted that the deformed and eroded layers of rock on the Earth's surface must have taken longer to accumulate than the few thousand years implied by the biblical creation story. Hutton estimated that the Earth's age needed to be measured in millions of years. It was not until the 20th century, with technology that could determine the rate of decay of radioactive elements, that the Earth's age was estimated at 4.6 billion years.

THE EARTH'S CORE

The Earth is layered inside. We know that the outer part is rocky. However, the Earth's mass is too great relative to its volume for it to be rocky throughout, which suggests that the composition of the core is very different.

The core is the densest part of the Earth and contains about one-third of its mass. It is mostly iron, with a few per cent of nickel and 10 per cent of a lighter element, which may be sulphur, oxygen, hydrogen or potassium. The core may reach as high as 5000°C. It has a crystalline inner part, with a pressure 3.6 million times greater than that at the surface, and a molten outer part. Turbulent flow in the molten region at a rate of a few centimetres per second generates the Earth's magnetic field. This geophysical phenomenon reaches above the atmosphere and protects the surface from the harmful effects of the solar wind and cosmic radiation.

00hr 00min 00sec
The birth of the universe, the Big Bang. Everything, including time, comes into existence.

16hr 38min 24sec
The solar system, including the Earth, is formed from the solar nebula.

20hr 48min 00sec
The first algae and primitive life forms appear in the Earth's oceans.

23hr 17min 57sec
The first land plants appear.

◄ 18–19 Formation of the planets

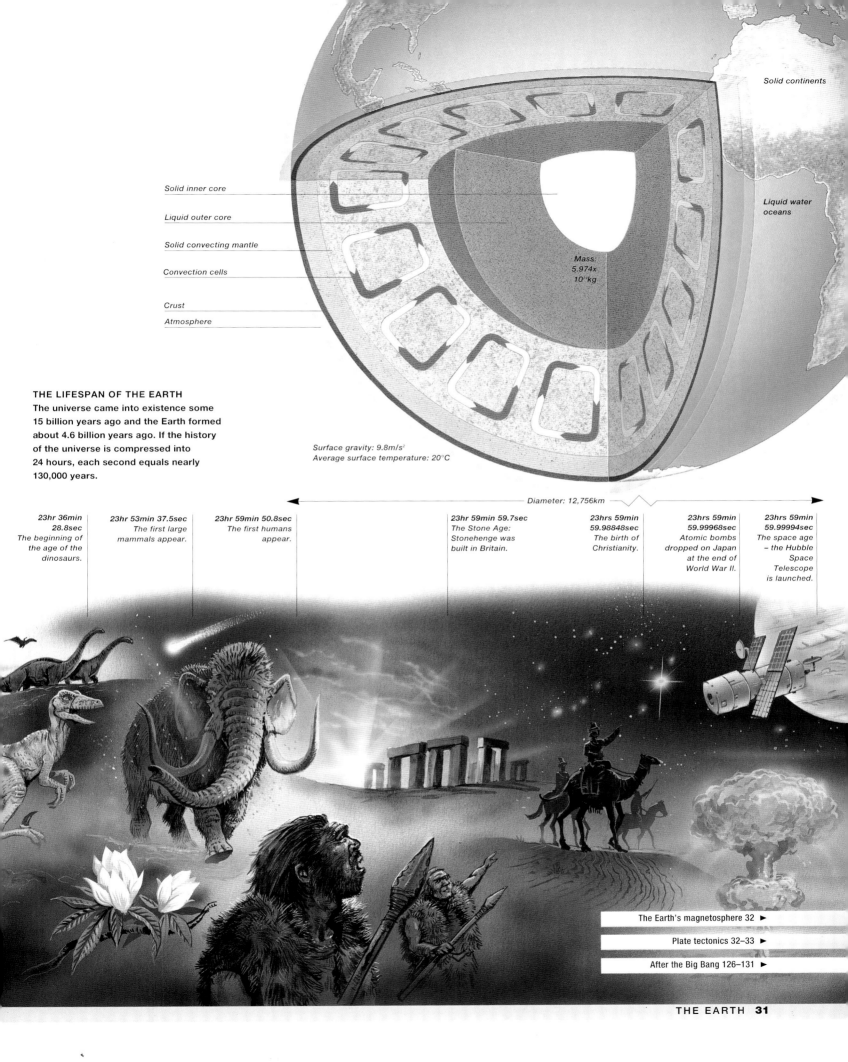

Solid continents

Liquid water
oceans

Solid inner core

Liquid outer core

Solid convecting mantle

Convection cells

Crust

Atmosphere

Mass:
5.974x
10²⁴kg

Surface gravity: 9.8m/s²
Average surface temperature: 20°C

THE LIFESPAN OF THE EARTH

The universe came into existence some
15 billion years ago and the Earth formed
about 4.6 billion years ago. If the history
of the universe is compressed into
24 hours, each second equals nearly
130,000 years.

Diameter: 12,756km

**23hr 36min
28.8sec**
*The beginning of
the age of the
dinosaurs.*

23hr 53min 37.5sec
*The first large
mammals appear.*

23hr 59min 50.8sec
*The first humans
appear.*

23hr 59min 59.7sec
*The Stone Age:
Stonehenge was
built in Britain.*

**23hrs 59min
59.98848sec**
*The birth of
Christianity.*

**23hrs 59min
59.99968sec**
*Atomic bombs
dropped on Japan
at the end of
World War II.*

**23hrs 59min
59.99994sec**
*The space age
– the Hubble
Space
Telescope
is launched.*

The Earth's magnetosphere 32 ▶

Plate tectonics 32–33 ▶

After the Big Bang 126–131 ▶

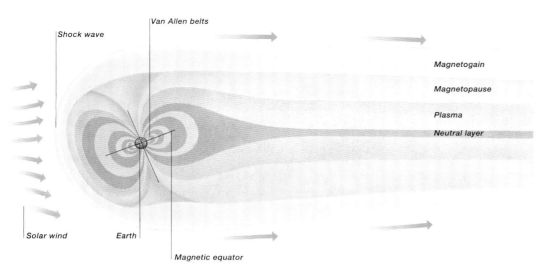

Shock wave

Van Allen belts

Magnetogain

Magnetopause

Plasma

Neutral layer

Solar wind

Earth

Magnetic equator

The crust owes its origin to the crystallization of molten rock that originated in the mantle and escaped upward. The oceanic crust that forms the floor of deep oceans is continually being created and destroyed by the processes of plate tectonics (see below). It is all less than 200 million years old and is about 6–11km thick. The continental crust forms the surface of the land and the floor of some shallow seas. It is much older, some parts are nearly four billion years old and varies from 25–90km in thickness. Most has been modified by processes such as erosion by wind, water and ice, and heating and remelting when continents collide.

It is a commonly held fallacy that if you could drill through the Earth's crust you would find a molten interior. Even below the rigid outer shell that forms part of the lithosphere, the convecting part of the mantle is essentially solid, and pockets of molten rock (magma) are rare.

MAGNETOSPHERE

The Earth's magnetic field is generated by motion in the fluid outer core, but its effects extend way out into space. Close to the Earth, the magnetic field is symmetrical. At greater distances its shape is distorted by the solar wind, which consists of a stream of charged particles from the Sun. The magnetosphere is the volume of space within which the Earth's magnetic field predominates. It is most extensive in the downwind direction (away from the Sun) and least extensive in the upwind direction (toward the Sun).

THE EARTH'S MANTLE AND CRUST

Above the core lie the 'rocky' parts of the Earth – the mantle and thin crust floating on it. In contrast to the core, the most abundant elements in the mantle are silicon, oxygen, magnesium, iron, aluminium and calcium. These elements are the fundamental constituents of the silicate minerals that make up most rocks. The rocks of the crust are again chemically distinct from the mantle; they are richer in silicon, aluminium, calcium, sodium and potassium.

The distinction between crust and mantle is important from a chemical point of view, but so far as physical processes go, the crust and the uppermost mantle behave as a single unit termed the lithosphere. This is a rigid shell about 100–200km thick. Beneath the lithosphere, the rest of the mantle is weaker and flows at a speed of a few centimetres a year. This flow takes place mostly in the form of convection currents that carry heat outward from the interior to the cooler regions nearer the surface.

PLATE TECTONICS

A special feature of the Earth's geology is that its outer shell (the lithosphere) is broken into a number of plates, each of which is moving at centimetres per year relative to its neighbours. This motion has three possible consequences and is the main cause of earthquakes and volcanoes.

Firstly, there is divergent plate movement where adjacent plates are pulled apart. This action draws material up from the deeper mantle to fill the space. The decrease in pressure experienced by the upwelling mantle makes it begin melting, and the resulting magma rises upward to form new oceanic crust.

The second type is convergent plate movement and occurs when two plates collide against each other. When the oceanic crust is on a separate plate to its adjacent continental plate, this convergence can result in the ocean floor being dragged underneath the continent, where it is eventually reabsorbed back into the deep mantle. An example of this is the Pacific coast of South America.

The Aurora Borealis or northern lights. Aurorae are caused by charged particles from the solar wind interacting with the Earth's atmosphere. They vary in colour from whitish green to deep red. This display shows a massive green arc and curtain.

◄ 20 The corona and solar wind

◄ 30 Internal structure of the Earth

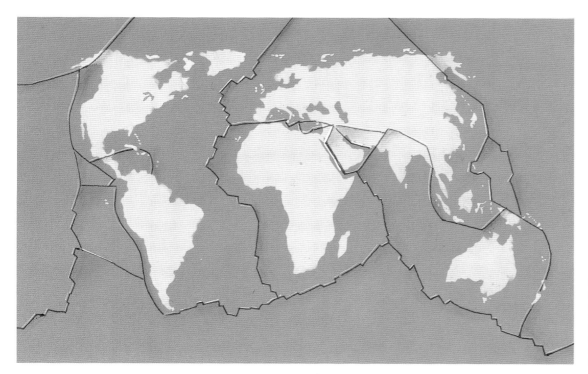

Below: A satellite image of California showing the stress patterns in the crust caused by the Pacific and North American plates grinding past each other. The density of the coloured bands indicates the severity of the stress, and these are noticeably focused on the fault lines (black).

Averaged out around the whole globe, the rate of this dual process of destruction and creation of new ocean floor is exactly balanced.

Sometimes, an ocean closes up completely, bringing two continents into collision. Continental crust, unlike oceanic crust, is too buoyant to be dragged far into the mantle and so colliding continents become buckled and thickened and eventually fuse together. This is how most mountains are formed and explains the existence of mountain belts such as the Alps and the Himalayas. By this stage, the motion that has been halted here will start to take effect at plate boundaries elsewhere.

The third type of plate movement is conservative, where the plates move parallel to each other, without large amounts of the Earth's crust being created or destroyed.

EARTHQUAKES

Most earthquakes happen near the boundaries between plates. They occur because rock masses that are in contact with each other do not slide freely past each other. Instead they stick and jerk free at irregular intervals when enough strain energy has been built up to overcome the resistance. This can cause powerful shaking at the Earth's surface, especially when the initial break point is shallow (less than about 10km deep).

The largest earthquakes tend to occur at convergent plate boundaries, where one plate is drawn down below the edge of another, and at conservative plate boundaries, where adjacent plates slide past each other. Examples of the former occur in Japan, Central and South America, and southern Europe. The most famous example of the latter is southern California: the San Andreas fault system allows a region of continental crust at the edge of the Pacific plate to slide north-eastward relative to the main North American continent at a rate of about a centimetre per year.

The eruption of the Kliuchevskoi volcano was photographed by astronauts aboard the space shuttle Endeavour in 1994. The eruption cloud reached over 18,000m above sea-level and the winds carried ash 1000km away.

◄ 32–33 Plate tectonics

VOLCANOES

Volcanoes mark sites where molten rock, known as magma, rises to the surface. They tend to occur along divergent or convergent plate boundaries (with some exceptions).

The most active volcanic zones are along divergent plate boundaries, where the plates are moving apart and magma is pulled up, creating new ocean floor. Averaged out over time, there are several eruptions per century along each stretch of divergent plate boundary.

The young, hot ocean floor is more buoyant than the older, cooler ocean floor to either side, and divergent plate boundaries are marked by volcanic ridges. A famous example is the Mid-Atlantic ridge, which separates the Eurasian and African plates on the east from the North and South American plates on the west. Like all such plate boundaries its crest lies mostly at 2–3km below sea-level, so most of its volcanic activity goes unnoticed. However, there is a notable exception in the form of Iceland where the rate of eruption over the past few million years has been sufficient to build the surface up to above sea-level.

The most obvious volcanoes lie in chains near the edges of the over-riding plates at convergent plate boundaries. The crust of the downgoing plate warms up as it is dragged

deeper and begins to melt. Furthermore, water driven out from the downgoing plate because of the progressively higher pressure as it goes down, passes upward into the overlying mantle and causes this to begin to melt as well. Magma rises upward to feed volcanoes that are usually about 70km above the top of the downgoing plate. The volcanoes in the Andes mountains along the western edge of South America are good examples of this type, as are the volcanic island arcs of the Caribbean, Indonesia and the western Pacific ocean.

In addition, gases contained within batches of magma will escape as they approach the surface. If the magma is of a particularly viscous kind, or if its gas content is especially great, the gas often escapes with such force that the eruption is explosive. This breaks the magma into glassy fragments described as volcanic ash, which can be driven upward in eruption columns that can reach 40km or more above a volcano in the most violent events. The ash cloud becomes dispersed downwind, so that towns and cities hundreds of kilometres away can become blanketed by it when it eventually reaches the ground. In other ash eruptions, a searingly hot dense cloud of ash may sweep across the ground at speeds in excess of 100km per hour.

More fluid or less gassy magma can reach the surface as a liquid, described as lava. Lava flows are less dangerous than ash flows because they travel more slowly, but will destroy anything that they encounter.

Not all volcanoes arise as a consequence of plate tectonics. Some isolated groups of volcanoes like Hawaii occur far away from plate boundaries. These are situated above pipe-like zones of upwelling in the mantle that may originate from as deep as the core–mantle boundary.

Magma rising to the surface does not always result in a volcano: it may lose so much heat to the surrounding crust that it solidifies, forming an 'igneous intrusion'. Examples up to tens of kilometres wide have been found.

RESURFACING THE EARTH

The surface of the Earth is very young because the crust is subject to processes that do not affect any of the other planets to the same extent. The oceanic crust usually survives for only 200 million years or less before being destroyed at a convergent plate boundary. Continents, on the other hand, are very old because they survive indefinitely without being dragged down into the mantle. However they bear few of the impact crater scars that characterize other planetary surfaces, because most of the continental surface is constantly being worn away by erosion (especially in mountain belts produced by continental collisions) or buried either by sediment from the eroding regions or by ash and lava erupted from volcanoes. The largest impact crater known on the Earth, the 180km diameter Chicxulub crater in Mexico, which was produced by a major impact 65 million years ago, is buried below several hundred metres of sediment.

The Earth above the Moon's surface. A photograph taken by an astronaut on board an Apollo spacecraft.

Collision course 88–89 ▶

ATMOSPHERE

Everyday human activities take place in the troposphere, which extends up to about 10km. Higher still, weather balloons fly into the stratosphere, which extends up to about 50km. Meteors – small amounts of dust from space – burn up in the lower thermosphere and the mesosphere.

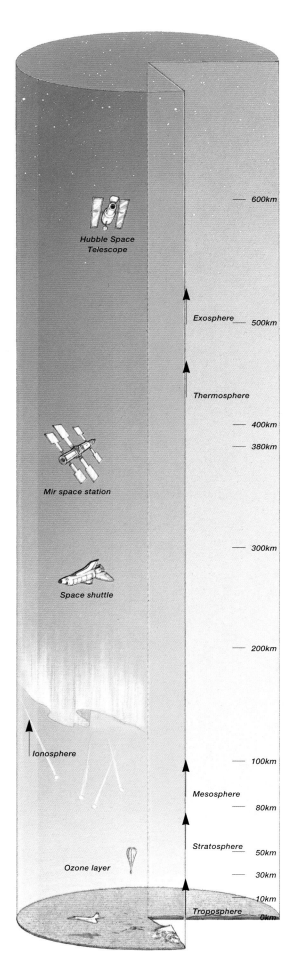

Hubble Space Telescope

Exosphere —— 600km

—— 500km

Thermosphere

—— 400km

—— 380km

Mir space station

Space shuttle

—— 300km

—— 200km

Ionosphere

—— 100km

Mesosphere

—— 80km

Stratosphere —— 50km

—— 30km

Ozone layer

—— 10km

Troposphere —— 0km

ATMOSPHERIC COMPOSITION

Gas	Percentage
Nitrogen	78
Oxygen	21
Water vapour	variable (up to 4)
Argon	0.9
Carbon dioxide	0.03

◄ 16–19 Formation of the solar system

◄ 26–27 Venus' atmosphere and acid rain

◄ 29 Greenhouse effect

◄ 32–33 Plate tectonics

THE OCEANS

Another reason the Earth is unique is its oceans of liquid water that extend over 70 per cent of its surface. These cover virtually all the oceanic crust to an average depth of 3.7km, and also drown the edges of most regions of continental crust. Oceans are known to have existed for nearly four billion years. Their water probably came from a mixture of steam escaping from the interior of the young Earth and ice delivered by impacting comets.

Microscopic plants (algae) floating in the sunlit upper reaches of today's oceans provide the basis of the food web that supports ocean life. It is the algae that give water its colour and that enable us to map the oceans' system of currents because different concentrations of algae are found in different water masses.

Many ocean currents take the form of giant circulatory systems, and they have an important influence on the climate of the countries whose coasts they wash. For example, the relatively mild climate of the British Isles and Scandinavia, compared to that of Labrador and Greenland (which are at the same latitudes at the opposite side of the Atlantic Ocean), is because of the Gulf Stream current that brings warm water from the region of Florida and Cuba. On the other hand, the famous fogs of San Francisco Bay are caused by a cold current flowing southward from Alaska.

Ocean currents are driven by the combined effects of the Earth's rotation and temperature differences between the poles and the equatorial regions. They are persistent features and should not be confused with tides, which alternate to and fro on a twice-daily basis.

THE EARTH'S ATMOSPHERE

The Earth's atmosphere, like that of Venus, originated by degassing from the planet's interior in processes such as volcanic eruptions. It probably began as a mixture of water vapour, carbon dioxide, sulphur dioxide and nitrogen, but its composition has been influenced by the development of life.

Some of the first primitive organisms on the Earth produced oxygen as a waste product, through the breakdown of carbon dioxide and sulphur dioxide. This paved the way for more familiar organisms (such as ourselves) that could not survive without oxygen to breathe and a high-altitude ozone layer to shield the surface from the harmful effects of ultraviolet sunlight. Moreover, without the ability of certain atmospheric gases to trap heat (the so-called greenhouse effect) the surface temperature would be about 30°C cooler.

Compared to the size of the Earth, the atmosphere is merely a thin layer. It is so sparse as to be almost unbreathable near the peaks of the highest mountains (8–9km above sea-level), the highest clouds occur at about 12km, and it is thin enough to offer virtually no resistance to artificial satellites orbiting as low as 200km above the surface.

Like ocean currents, the atmosphere is in circulation, driven by the temperature difference between the equator and the poles. The flow is strongly influenced by the Earth's rotation, and also changes seasonally. Although the general pattern of circulation is understood, there are many smaller, random effects that make it impossible to produce reliable weather forecasts more than a few days ahead.

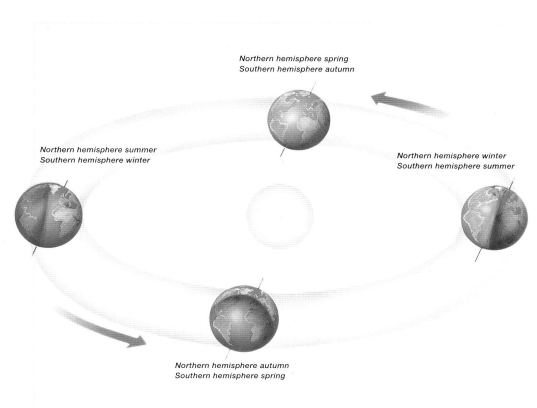

Northern hemisphere spring
Southern hemisphere autumn

Northern hemisphere summer
Southern hemisphere winter

Northern hemisphere winter
Southern hemisphere summer

Northern hemisphere autumn
Southern hemisphere spring

SEASONS AND THE EARTH'S ORBIT

The rotation of the Earth about its spin axis is responsible for day and night, while the Earth's orbit round the Sun takes a year.

If the Earth's spin axis were exactly upright relative to the plane of its orbit round the Sun (ie, at 90° to it) there would be no seasons. However, it is inclined at just over 23°, and points to the same direction in space through the year. Therefore, for part of the orbit the north pole is inclined toward the Sun, so that the northern hemisphere receives more sunlight and has longer hours of daylight than the southern hemisphere. During this part of the orbit it is summer in the northern hemisphere and winter in the southern hemisphere.

At the opposite side of the orbit the north pole is inclined away from the Sun so it is winter in the north but summer in the south. Twice a year (at the equinoxes) the inclined spin axis lies exactly side-on to the Sun, so that day and night are equal in both hemispheres.

Although the orientation of the spin axis can be regarded as fixed over a single year, it does in fact have a very slow wobble taking 25,800 years to complete, which is responsible for the precession of the equinoxes.

HUMANS AND THE ECOSPHERE

The interrelated environments in which life exists form a fragile system known as the ecosphere. There are many examples in the geological record of climate fluctuations, when the globe was warmer or colder than today. In the past these have been triggered by sudden natural events, such as asteroid impacts and large volcanic eruptions, or set off gradually by such factors as the changing distribution of continents, variations in the amount of sunlight caused by orbital effects, or the influence of life itself. These changes, especially the rapid ones, have often been associated with the disappearance of particular environments and the extinction of plants and animals, such as dinosaurs.

Modern human activities are forcing change at rapid rates. The burning of fossil fuels (coal and oil) has caused a substantial increase in the amount of carbon dioxide in the atmosphere during the second half of the 20th century. Carbon dioxide is one of those gases responsible for the greenhouse effect, and the increase in carbon dioxide is blamed for a global average temperature increase of half a degree during that time. Unless the rate of carbon dioxide emission is brought under control, the climate could be more than 2°C warmer a hundred years from now. The effects of this on local weather systems are hard to predict, but one certain consequence is a rise in global sea-level of up to half a metre caused by melting of ice in Greenland and Antarctica and thermal expansion of seawater in general as it warms up.

Other human influences are seen in the depletion of the ozone layer, notably in the 'ozone hole' over Antarctica, caused by the release of gases called chlorofluorocarbons (CFCs), and in the felling of enormous tracts of vital tropical rain forest, which absorb carbon dioxide and emit oxygen.

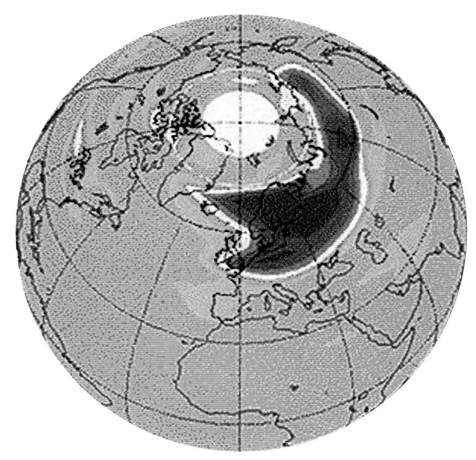

The ozone is vital as it blocks the Sun's harmful radiation. Man-made chlorofluorocarbons (CFCs) destroy the ozone layer. This image shows the hole in the northern hemisphere on January 11, 1992. The strength of chlorine monoxide, the main ozone-destroying molecule, is shown in false colour, increasing from blue to green, yellow, red and dark purple.

Global warming 38–39 ►

Tides 45 ►

Collision course 88–89 ►

GLOBAL WARMING

Global warming is a rapid climate change resulting from human activities on the Earth's atmosphere. As a geologist I'm used to thinking about climate change: most rocks I see on field trips were formed when the Earth's average temperature was warmer or colder than today. I'm used to studying places with extreme climates that have undergone major changes, but this scares me.

BY DR DAVID ROTHERY

Above: Smoke from rainforest burning in Borneo caused a pall of smoke half the size of North America over South-East Asia for several months in late 1997.

" Why should the prospect of present day climate change on Earth fill me with dread? There are two reasons. The first is the speed at which things are happening. Industrial growth, and in particular carbon dioxide produced by burning fossil fuels, caused the global average temperature to increase by about 0.5°C over the 140 years following 1860. By the year 2100 the amount of carbon dioxide in the atmosphere will be double the pre-industrial level even if we could stabilize carbon dioxide emissions at the present level, so the temperature seems set to rise farther by 1–3.5°C during the next 100 years. That's a much faster rate of change than anything within the past million years, even when the ice caps were in advance or full retreat as the globe see-sawed between interglacial and glacial periods. The ecosystem is used to climate change, and redistribution or loss of species is a natural process, but no-one knows if the globe can sustain such a large and rapid change as the one we are causing now without severe repercussions.

What is the greenhouse effect? The Earth's surface is heated by light from the Sun. This heat is radiated back from the surface as infrared radiation, which would escape out to space if we had no atmosphere, but water vapour and carbon dioxide in the atmosphere prevent this occurring, so the global temperature increases. It is thanks to this effect that the Earth is a comfortable place to live: without it the Earth would be at least 30°C colder. Our problem today is an enhanced greenhouse effect that occurs mainly because the burning of fossil fuels has added extra carbon dioxide to the atmosphere, but it is made worse by industrially or agriculturally produced gases such as methane and nitrogen dioxide.

An example of what can happen when the global climate is knocked rapidly off course lies in the events of 65 million years ago when a large asteroid hit the Earth. The local devastation was immense, but that's not the point. So much dust

No-one knows whether the globe can sustain such a large and rapid change as the one we are causing now without severe repercussions.

was thrown into the atmosphere that the sunlight reaching the ground was greatly dimmed and the globe became much cooler than normal. Plants died, and so did the animals that depended on them, directly or indirectly, for food. Within a few years, as the air cleared, plants growing from seeds and spores were able to re-establish themselves, but something like half the species of land and marine animals (including the last of the dinosaurs) became extinct. We do not know how much this was an immediate effect of starvation, or a slower (100,000 years) process resulting from the disturbance to the ecosystem. Global warming caused by human activity is unlikely to have quite such devastating consequences, but the dangers of pushing the climate too far and too fast are clear.

My other reason for worry is that modern societies and the global economy do not respond easily to changes of the sort that will happen if global warming continues. Ice Age cavemen probably thought little of relocating themselves north or south to follow the game on which they depended. We, however, live in a world with political borders and ownership of property. If a poor country loses most of its agricultural land because the climate has changed, its neighbours are not likely to welcome tens of millions of refugees, particularly if they already have similar problems of their own.

It's not just a matter of the climate becoming locally warmer or colder, wetter or drier, stormier or less stormier. As global average temperature rises, so does sea-level because of the thermal expansion of seawater and melting of the ice caps. Sea-level has risen by over 10cm in the past 100 years, and looks likely to rise about a further 50cm before the year 2100. The Netherlands can afford to build its sea defences higher, but can the government of Bangladesh? And what about the small island states like the Maldives and Kiribati? They have no high ground to retreat to, and already their fresh water supplies are under threat.

Sometimes people confuse global warming with an effect called El Niño. This is a local climate fluctuation centred in the equatorial Pacific that is triggered by a reversal in the direction of the surface current. During an El Niño event there may be drought in the region of Australia and excessive rainfall on the west coast of America. El Niño is nothing to do with global warming, although its effects may become more extreme as the global temperature rises, and it is again the poorest nations that may suffer the worst effects.

Don't get the impression I lose sleep over global warming. I'm used to it. The possible consequences of climate change are just another aspect of modern life. I'm pleased that global warming is now firmly on the political agenda to the extent that countries send high-profile delegations to United Nations conferences about limiting the emissions of greenhouse gases. However the cynic in me sees that the cuts that are agreed to are the bare minimum necessary to gain political favour in the short term, at the expense of the medium-term benefits that would accrue from making bigger cuts. The trouble is that an industrial nation is likely to feel economically disadvantaged if it makes a bigger cut than its competitors, and a developing nation feels it should have the right to do as much damage to the environment per capita as the developed nations have done.

On an individual level there may seem nothing you or I can do about this, but actually there is. The world is made up of individuals. If enough of us moderated our lifestyles, we could significantly reduce the rate of global warming, so that not just the ecology but also human societies would have a better chance of coping with it. 🙶

The global average temperature has risen by about 0.5°C over the last 140 years. It could rise by as much as 3.5°C over the next 100 years. As well as making the ice caps melt and sea-water levels rise causing severe flooding in many coastal areas, global warming will affect inland areas and may alter weather patterns everywhere.

THE MOON

THE MOON, A FAMILIAR BRIGHT FEATURE IN OUR
NIGHT SKY, IS IN FACT A ROCKY BODY WITH NO LIGHT
SOURCE THAT SHINES ONLY BY REFLECTED SUNLIGHT.

Waxing crescent at 3 days   **Waxing crescent at 6 days**

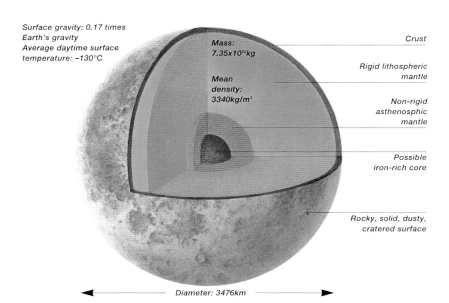

Surface gravity: 0.17 times
Earth's gravity
Average daytime surface
temperature: −130°C

Mass:
7.35x10²²kg

Mean
density:
3340kg/m³

Crust

Rigid lithospheric
mantle

Non-rigid
asthenosphic
mantle

Possible
iron-rich core

Rocky, solid, dusty,
cratered surface

Diameter: 3476km

VITAL STATISTICS

A: 356,410km
B: 406,700km
C: orbits in 27.3 Earth days
D: rotates in 1 Earth day

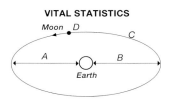

Earth   Moon

The Moon's diameter is 0.27
times the size of the Earth's

It is a strange coincidence that the Moon and Sun appear the same size in the sky. This is, perhaps, why most early civilizations did not recognize the fundamental difference between the Sun and the Moon. We now know that the Moon is a dark body, which shines only in light reflected from the Sun. When the Moon is opposite the Sun in our sky, we see its whole face illuminated (full moon), but when close to the Sun in our sky its unlit hemisphere faces us and the Moon is invisible (new moon). In between these positions we see a partly illuminated hemisphere of the Moon.

THE MOON: A SMALL PLANET

In many respects, the Moon can be regarded as the smallest of the terrestrial planets. It is unusual in that it orbits another planet (the Earth) rather than going directly round the Sun. The Moon is the only body other than the Earth that has been visited by humans, and from which we have collected samples (382kg in all). It has also been visited by a considerable number of spacecraft.

Analysis of the lunar samples suggests that the Moon was probably formed from the remnant of a Mars-sized body that collided with the juvenile Earth in a giant impact about 4.5 billion years ago. Any iron-rich core that this body had appears to have been absorbed into the Earth's core, whereas the Moon grew out of the mostly rocky debris thrown into space by the impact.

INSIDE THE MOON

Just like the Earth, the Moon is known to have an internally layered structure. The limited data we have about the interior comes from instruments installed on the surface by the Apollo missions, which detected moonquakes and measured the vibrations. These findings suggest that if it

does have an iron-rich core, it must be less than 360km wide and make up only two per cent of the Moon's mass. The lack of a large, fluid core explains the almost total lack of a lunar magnetic field, which is only one ten-millionth of the strength of the Earth's magnetic field.

Whatever core the Moon has is overlain by a rocky mantle, surfaced by a crust of slightly different composition. This is about 60km thick on the side nearest to Earth and about 100km thick on the far side. The Moon's rigid outer shell (lithosphere) is about 1000km thick, and only the very deepest mantle is hot enough to be flowing convectively.

This contrast with the Earth, in which all but the very uppermost mantle is convecting, is because the Moon's small size means that its surface area is much greater relative to its volume and therefore more heat has been able to escape from its surface. This means that the Moon has cooled down much more since its origin than the Earth has been able to.

THE LUNAR SURFACE

The pattern of dark patches on the Moon's surface is familiar to everyone. Some people recognize these as the face of the 'Man in the Moon'; others see a different picture. However, the fact that the same pattern is always there shows that we always see the same side of the Moon. This is because the Moon rotates on its axis exactly once per orbit.

The Moon's dark patches are now known to be low-lying regions that were flooded by vast outpourings of basaltic lava, known as the lunar maria (Latin for 'seas', singular mare). Radiometric dating of samples brought back to the Earth has enabled scientists to date this episode at between 3.0 and 3.9 billion years ago. The maria are not seas in any familiar sense, and appear never to have contained any water.

The remainder of the Moon's surface is made up of ancient highland crust, made of a rock type known as anorthosite. Samples have enabled the highlands to be dated at a little more than four billion years old.

THE MOON'S CRATERED SURFACE

The other most noticeable features on the Moon's surface are craters. Once presumed to be volcanic in origin, it was only with the detailed observations made possible by the Apollo missions that they became accepted as impact craters, caused by the impact of comets and asteroids onto the Moon's surface. These

First quarter at 7 days   Waxing gibbous at 11 days   Full moon at 15 days   Waxing gibbous at 19 days   Waxing crescent at 26 days

collisions typically occur at speeds of several tens of kilometres per second. A crater about 30 times the diameter of the impacting body is excavated by a shock wave generated at the point of impact. This always produces a circular crater, like a pebble falling into sand, unless the angle of impact is extremely oblique.

The Moon has craters of various sizes. Most of those more than a few hundred kilometres across have been flooded by lava, but there are many craters about 100km wide that are easy to see in binoculars. In fact, the better

The southern part of Mare Imbrium looking south to the Montes Carpatus taken from Apollo 17. The crater in the middle is Pytheus, 20km in diameter with a sharp rim and hilly floor. Crater Copernicus can just be seen at the top.

the detail in which you view the Moon, the more craters you find. There are two reasons for this. Firstly, the Moon's lack of atmosphere means its surface is unprotected from potential impactors. An atmospheric layer around a planet helps to burn up any object approaching the surface. Secondly, unlike the Earth, the Moon has no atmosphere or active geological processes through which old craters would be worn away: the only process on the Moon that can destroy a crater is the formation of a new crater on top of it.

The low-lying maria have fewer craters than the highlands because they are younger, which means there has been less time for craters to form there. However, they are not that much younger than the highlands and this proves that the rate of cratering declined about four billion years ago.

PHASES OF THE MOON
The interval between one new moon and the next is 29.53 days.

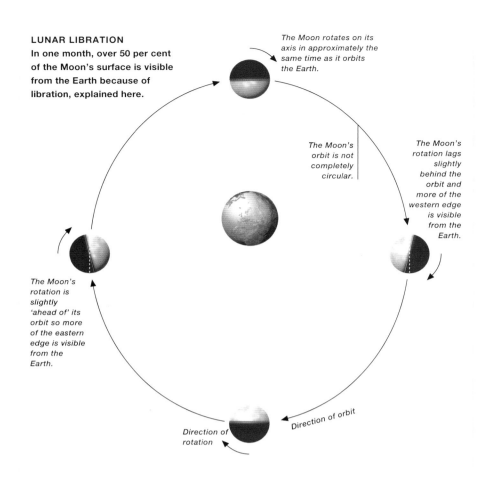

LUNAR LIBRATION

In one month, over 50 per cent of the Moon's surface is visible from the Earth because of libration, explained here.

The Moon rotates on its axis in approximately the same time as it orbits the Earth.

The Moon's orbit is not completely circular.

The Moon's rotation lags slightly behind the orbit and more of the western edge is visible from the Earth.

The Moon's rotation is slightly 'ahead of' its orbit so more of the eastern edge is visible from the Earth.

Direction of rotation

Direction of orbit

The Moon is thus our main source of evidence for what was going on in our part of the solar system at that time. Any evidence left on the Earth has long since been worn away by erosion and deformation. The Moon gives us an idea of how the impact rate is changing with time and helps us to predict future impactors that might hit the Earth.

The cratering time scale established by counting the numbers of craters on regions of the Moon, which we have been able to date by other means, provides the best means of estimating the ages of areas of other cratered bodies in the inner solar system. However, this does not work in the asteroid belt and beyond because there is evidence that the nature and number of impacting bodies were different in those regions.

The Apollo astronauts found very few areas of bedrock (solid rock) because the Moon is so heavily cratered, Instead they encountered a lunar soil or 'regolith' consisting of fragments flung out of craters by impacts, and dotted with larger blocks flung out from larger craters.

THE MOON'S ORBIT

The Moon orbits the Earth so closely that there are significant tidal forces between the two bodies. The Moon's influence on the Earth is most evident in the Earth's ocean tides. The Earth's effect on the Moon is even greater, and tidal forces have slowed the Moon's rotation down so that it rotates on its axis only once per orbit. This explains why we always see the same side of the Moon. This effect is known as synchronous rotation.

Above: The Moon rotates on its axis in approximately the same time as the Moon orbits the Earth. This means it always shows the same hemisphere to the Earth. Because the Moon's orbit is not completely circular, and it is inclined slightly to the Earth, it is possible to see more than 50 per cent of the Moon during one month. This effect is known as libration. Over a 30-year period as much as 59 per cent of the Moon's surface can be seen, and more of the Moon's north and south polar regions can be seen at different times. The Moon's phases are not shown here.

Right: Because the Moon's surface has endured billions of years of impacts it consists mainly of fragments of rock with large boulders thrown out from larger craters.

◄ 40–42 The Moon's cratered surface

◄ 18–19 Formation of the planets

In fact, because the Moon rotates at a constant rate but moves in a slightly elliptical orbit and therefore at a variable speed, the rotation is sometimes slightly ahead and at other times slightly behind its average position seen from Earth. The Moon thus appears to wobble slightly, so that although most of the far side is always invisible from Earth, nearly 60 per cent of the Moon's surface can be seen by observing it at different times.

The best way to see this for yourself is to make careful note of features near the edge of the lunar disk around the time of successive full moons. This effect is known as libration.

FUTURE LUNAR EXPLORATION

There is still much to be learned about the geological history of the Moon, and there are several low-cost lunar exploration probes planned by the American, Japanese and European space agencies. There will probably be human bases on the Moon by the year 2050. How self-sufficient they could be in fuel, food and mineral resources remains to be seen, but a lunar base would certainly provide valuable experience before beginning the more serious undertaking of colonizing Mars.

The Lunar Prospector Mission has revealed the existence of water ice near the poles, making them a possible suitable location for the first bases on the Moon. In addition, the cold polar environment would provide an excellent site in which to position infrared telescopes.

MISSIONS TO THE MOON

Name	Date	Aims and achievements
Lunik 2	1959	First human artefact to reach the Moon.
Lunik 3	1959	First pictures of the far side of the Moon.
Rangers 7, 8 & 9	1964–5	Deliberate crash landings on Moon. Obtained first close-up pictures of Moon showing boulders and metre-sized craters.
Luna 9	1966	First soft landing in Oceanus Procellarum.
Luna 10	1966	First probe to orbit the Moon.
Lunar Orbiters	1966–7	Photography of possible landing sites.
Apollo 8	1968	First manned orbits of the Moon.
Apollo 11	1969	First manned landing at Mare Tranquillitatis.
Apollo 12	1969	Manned landing and explo ration. Oceanus Procellarum.
Luna 16	1970	Collection and return of samples from Mare Fecunditatis.
Luna 17	1970	Lunokhod 1 rover travelled 20km on lunar surface in Mare Imbrium.
Apollo 14	1971	Manned landing and exploration in Fra Mauro.
Apollo 15	1971	Manned landing, exploration near Hadley Rille with lunar rover.
Apollo 16	1972	Manned landing and exploration at Cayley-Descartes.
Apollo 17	1972	Last Apollo landing, south-eastern rim of Mare Serenitatis.
Clementine	1994	Multispectral imaging and altimetry from orbit.
Lunar Prospector	1997	Detailed mapping of surface composition, gravity and magnetic fields from polar orbit.

Mir Space Station photographed from the shuttle Atlantis after separation. This was the first of the Mir-Shuttle link-ups between 1995 and 1998. These joint operations were designed to pave the way for assembly of the International Space Station. Any future moon base would also need to be an international effort as it would be too expensive for one nation to fund.

Tides 45 ►

Collision course 88–89 ►

Orbits 90 ►

Spaceflight history 146–147 ►

Robotic missions 148–149 ►

Observing the Moon 160 ►

If the Moon's orbit lay exactly in the plane of the Earth's orbit, the Moon would pass directly in front of the Sun at the time of each new moon, and into the Earth's shadow at full moon. In fact, the Moon's orbit is inclined at 5°, so these events (known as eclipses) occur only when the Moon happens to be passing through the Earth's orbital plane exactly at the time of new or full moon.

The passage of the Moon in front of the Sun is described as a solar eclipse. When the Moon is near the closest point of its orbit, it is just big enough to obscure the Sun's disk completely (a total solar eclipse). For those people within the track of the Moon's shadow, never more than 270km wide, the sky goes dark, enabling them to see the Sun's outer atmosphere, the corona. Because the Moon's shadow is so narrow, it is more common to see a partial solar eclipse when just part of the Sun's disk is hidden by the (invisible) Moon.

The Earth casts a much bigger shadow, and it is more common to see a lunar eclipse, when the Moon passes into the Earth's shadow. The Moon does not usually disappear from view as you might expect, because refraction in the Earth's atmosphere bends some sunlight round to the Moon, like light through a prism, bathing it in a dim red glow.

Above: A total lunar eclipse. These occur at full moon when the Moon passes into the darkness of the Earth's shadow. The Moon appears coppery red because of sunlight refracted through the Earth's atmosphere.

Right: The diamond ring effect of a total solar eclipse. As the Moon is not smooth but composed of mountains and valleys, part of the Sun can shine through a lunar valley just before or after a total eclipse. Some mountains can be seen to the right and at the bottom.

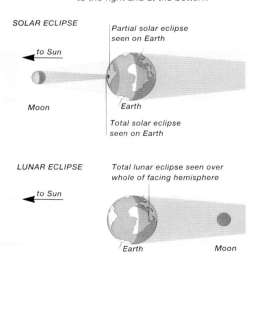

SOLAR ECLIPSE

to Sun

Partial solar eclipse seen on Earth

Moon   Earth

Total solar eclipse seen on Earth

LUNAR ECLIPSE

to Sun

Total lunar eclipse seen over whole of facing hemisphere

Earth   Moon

◄ 40–41 Phases of the Moon

◄ 21 The corona and solar wind

WHAT CAUSES TIDES

Ocean tides are the result of the gravitational pull of the Moon and the Sun. Although the Moon is much smaller than the Sun, it is also much closer to the Earth so its effect is about twice that of the Sun. When the Moon and Sun are in roughly the same direction (about the time of new moon), they each pull the oceans on the near side of the Earth towards them. They also pull the solid Earth towards them away from the oceans on the far side of the Earth.

The effect is to produce two tidal bulges in the ocean on opposite sides of the Earth. These tidal bulges will not rotate with the Earth, but will stay lined up with the bodies that produced them, bringing about two high tides a day. The effect is exactly the same if the Moon is on the opposite side of the Earth to the Sun (at the time of full moon). At full moon and new moon, the height of the tides is greatest (spring tides).

When the Sun, Moon and Earth are not lined up, the situation is more complicated. Tides are least strong when the Sun and Moon are at right angles to each other in the sky (when the Moon is at first and last quarter). In this situation the solar tide and lunar tides compete against each other. The lunar tide usually wins,

but the difference between high and low tides is much less. These are described as neap tides.

In the open oceans the tidal range is mostly less than a metre, but water piles up on many coastlines to give ranges of several metres. The record is the bay of Fundy, Nova Scotia, where tides have a range of 15m.

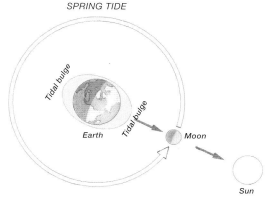

SPRING TIDE

Tidal bulge · Earth · Tidal bulge · Moon · Sun

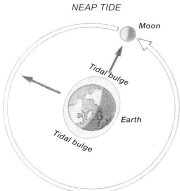

NEAP TIDE

Moon · Tidal bulge · Earth · Tidal bulge

The total eclipse of the Sun occurs when the Moon's disk completely covers the Sun's surface. When the light of the Sun's photospere is hidden, the ethereal light of the corona, the outer atmosphere of the Sun, can be seen.

Gravitation 90–93 ▶

Light 94–95 ▶

MARS

MARS IS OUR NEAREST PLANETARY NEIGHBOUR AND CAN BE OBSERVED WITH EVEN SMALL TELESCOPES. ITS RED COLOUR IS OBVIOUS WITH THE NAKED EYE.

Surface gravity: 0.38 times Earth's gravity
Average surface temperature: –60°C

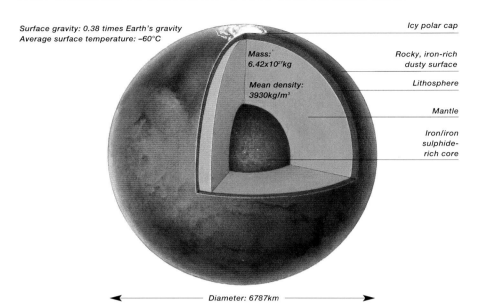

Icy polar cap

Rocky, iron-rich dusty surface

Lithosphere

Mantle

Iron/iron sulphide-rich core

Mass:
6.42×10^{27}kg

Mean density:
3930kg/m³

Diameter: 6787km

VITAL STATISTICS

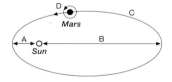

A: 207,000,000km
B: 249,000,000km
C: orbits in 686.98 Earth days
D: rotates in 24h 37min 23sec

Mars' diameter is 0.53 times the size of the Earth's

Angle of equator to ecliptic: 23.98°
Angle of orbit to ecliptic: 1.85°

Known satellites: two

The red colour of Mars is because of the iron-rich compounds in the rock and dust on its surface. Mars is close enough to the Earth for even small telescopes to pick out features, such as dust storms and the polar ice caps.

COMPOSITION OF MARS

Observations suggest that Mars contains an iron-rich core, about 1700km in diameter. The low density of Mars compared to the other terrestrial planets hints that this core may also contain a significant amount of sulphur. Apparently, this core is not convecting enough to create as strong a magnetic field as the Earth. Indeed, it was not until 1997 that Mars Global Surveyor detected the weak and patchy magnetic field of Mars.

The core is surrounded by a molten rocky mantle somewhat denser and perhaps some three times as rich in iron oxide as that of the Earth, overlain by a thin crust. The lack of plate tectonics and absence of current volcanic activity on Mars implies this mantle is also non-convecting. One massive surface feature, the 4500km long Valles Marineris, may be a fracture in the Martian crust caused by internal stresses, but scientists do not know.

MARTIAN SEASONS

Like the Earth, Mars experiences seasons. Its orbit is not circular and it is much closer to the Sun during the southern summer than in the northern summer. So southern Martian summers are warmer than northern ones, but because the planet moves faster when it is closer to the Sun they are shorter and southern winters longer and colder than those in the north. One result of this is that the southern residual cap retains some frozen carbon dioxide (which melts at a lower temperature) as well as water.

Studies have shown that both polar caps are composed of a permanent layer of mostly water ice, with a seasonal cover of carbon dioxide that condenses out of the thin atmosphere in the autumn and dissipates in the spring. As it condenses, the carbon dioxide forms temporary clouds, which hover over the pole that is about to enter winter.

DUST STORMS

Scientists have discovered that Mars has seasonal dust storms that can grow to planetwide proportions. When Mariner 9 arrived at Mars in 1971, it found the planet enveloped in a global dust storm and the first pictures sent back to Earth were completely blank.

These dust storms are caused by the interaction between the Sun heating the dry atmosphere and the seasonal changes of the polar caps setting up convection currents. These Martian winds may gust up to 90km/h at the surface. Evidence for prolonged cycles of erosion and deposition lies in the layered deposits that surround the poles. Surrounding this laminated terrain are broad belts of dunes, which reveal the wind action.

DRY AND COLD

Mars is smaller and less massive than the Earth or Venus, and so has a lower surface gravity. This means that Mars cannot hold on to a dense atmosphere. In addition, Mars' lower volume means that it could not generate and retain the same amount of internal heat as Venus or the Earth, and does not maintain the same level of volcanic activity.

The carbon dioxide atmosphere of Mars must have gradually been depleted with time, and as surface pressures and temperatures decreased, the water froze out and became locked as subsurface ice and in the polar caps.

In the late 1960s the Mariner 4, 6 and 7 spacecraft confirmed that the surface resides under only a thin atmosphere of carbon dioxide, with a pressure of only one hundredth of that at the Earth's surface at most, and in places even lower.

They also revealed that Mars is cold, with mean annual temperatures ranging from –58°C at the equator to –123°C at the poles. At these temperatures and low pressures liquid water cannot currently exist on the Martian surface, although the Mariner and subsequent Viking pictures revealed evidence for the ancient action of flowing water.

SURFACE FEATURES ON MARS

Mars can be divided into two regions, the heavily cratered plateau of the southern highlands, and the low-lying plains of the northern hemisphere. These regions are divided by a highly irregular strip of eroded, broken terrain where the surface drops 1–3km onto the northern plains. This marked asymmetry in topography is called the Martian dichotomy, and its origin remains a mystery. The cratered highlands represent the ancient crust of Mars, recording an impact bombardment that ended about 3.8 billion years ago. At first glance these highlands appear very Moon-like, but the Martian craters have been highly eroded and filled in by later geological activity. The northern plains are less heavily cratered, and may largely be made up of lava flows, as well as sediments deposited by wind and water.

Another major topographic feature of Mars is the Tharsis Bulge which rises 10km above the mean surface level of the planet and measures 4000km across. This enormous structure sits astride the Martian dichotomy and is capped by several giant volcanoes including Olympus Mons.

MARTIAN VOLCANOES

Olympus Mons is the largest volcano in the solar system. Its peak rises to a staggering 27km above the mean surface level of Mars. More than three times as high as Earth's Mount Everest, it has a diameter of some 520km (if dropped on top of them it would virtually cover Utah or Poland). Olympus Mons is surrounded by a huge cliff up to 6km high, where the lower flanks appear to have fallen away in a gigantic landslide. This collapse may have generated the peculiar blocky terrain of ridges separated by flat areas, the Olympus Mons aureole, that extends from the base of the cliff up to 1000km from the volcano's summit.

Olympus Mons' summit contains a nested set of volcanic craters (*calderae*), the largest of which is 80km across. Each caldera was formed when the solid top of the volcano collapsed as the underlying magma drained back into the mantle. The last volcanic event formed a pit 3km deep.

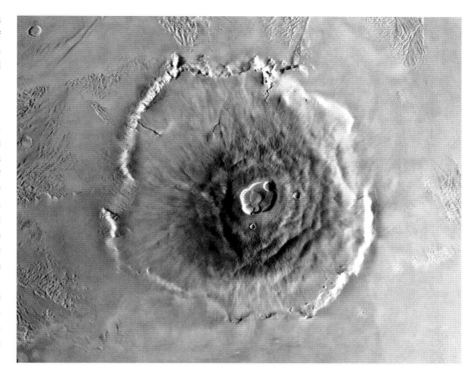

Above: A mosaic of the Olympus Mons volcano on Mars created from images taken by the Viking 1 orbiter in 1978. Olympus Mons is the largest volcano on Mars and possibly the largest in the solar system. It is about 500km across and 27km high. Scientists estimate that it is about 200 million years old.

Left: A mosaic of the Schiaparelli hemisphere of Mars created from images taken by the Viking orbiter in 1980. The large crater near the centre is Schiaparelli, which is about 500km in diameter. The dark streaks are caused by erosion and/or darker dust deposited by the strong winds on Mars and the white areas to the south are carbon dioxide frost.

Water on Mars 49–50 ▶

PERCIVAL LOWELL

Percival Lowell (1855–1916) was a wealthy, well-connected member of Boston's scientific circles. He was fascinated by Mars and, when approached by a Harvard astronomer, W H Pickering, for help in funding an observatory in Arizona to view the 1894 opposition of Mars, Lowell funded the entire project. He became director of his own observatory in Flagstaff and oversaw its work as a benign dictator. His staff had to undertake tasks he wanted, and were not to question his judgement. A E Douglass was fired in 1901 for criticizing Lowell's conclusions about there being canals on Mars. Lowell's fascination for Mars led to his publishing detailed maps of networks of up to 160 canals. He also popularized his theories in a series of books that included *Mars and its Canals* (1906) and *Mars as an Abode of Life* (1910).

A mosaic of the Valles Marineris hemisphere of Mars created from images taken by the Viking 1 orbiter spacecraft in 1980. The Valles Marineris canyon system is over 4500km long and up to 7km deep, with individual canyons up to 200km wide. Two of the three Tharsis volcanoes, each about 25km high, are visible on the left of the hemisphere.

There are three other huge volcanoes on Tharsis: Ascraeus, Pavonis and Arsia Mons. They are all several hundred kilometres across, and the highest of them (Ascraeus Mons) rises to 26km above the mean surface level of Mars.

Like Olympus Mons, they are shield volcanoes, formed by long-term eruption of copious amounts of fluid basaltic lava. These Martian volcanoes have grown so enormous because they have sat over their molten source for hundreds of millions of years, whereas volcanoes on Earth get moved off their magma supplies more rapidly by plate tectonic motion. There are also a number of smaller volcanoes in Tharsis, as well as elsewhere on Mars.

THE VALLES MARINERIS

One of the most spectacular features in the solar system is the giant canyon system of Valles Marineris, which is named in honour of the NASA Mariner 9 (1971) spacecraft that discovered it. This deep gash in the crust of Mars extends for 4500km eastward from the Tharsis Bulge, plummets to depths of 7km below its surroundings, and is up to 200km wide. On Earth, it would stretch from San Francisco to Boston, and even its minor tributaries would be the size of the Grand Canyon. Unlike the Grand Canyon, however, the Valles Marineris was not carved by running water, but by the rifting and extension of the Martian crust as the planet cooled. Nevertheless, water may have contributed to the subsequent erosion and widening of Marineris.

WATER ON MARS

The ground is permanently frozen to a depth of several kilometres on Mars and there is a great deal of frozen water locked up in the polar caps, and probably large amounts at depth beneath the surface.

Despite the absence of any surface water today, the Mariner and Viking pictures show that the ancient cratered highlands are laced with intricate networks of valleys. Many of these valleys only superficially resemble the river systems of Earth, and many may have been formed by ground collapse caused by the gradual loss of subsurface ice or water. Yet if some of these valleys were cut by running water, this must mean that a few billion years ago Mars must have had a denser atmosphere and a warmer, wetter climate a few billion years ago.

The most compelling evidence for the existence of running water on Mars in the past is the features that resemble outflow channels. The largest of these are located around the plain of Chryse Planitia, including Ares Vallis – the landing site of Mars Pathfinder. The channels emerge from areas of chaotic blocky terrain at the edge of the highlands, and drain down into the northern lowlands. They are huge and can be up to 200km wide and 2000km long.

Analyses of the images of these channels suggest that they were formed by catastrophic floods that burst from the ground and formed the chaotic terrain. Perhaps volcanic heating or marsquakes caused underground reservoirs of water to be suddenly released onto the surface. Astronomers have calculated that the water in these channels may have flowed at a peak rate of 1km^3 per second. Such catastrophic floods can occur in any climate, and the evidence suggests that the outflow channels occurred at many different times in Martian history.

VANISHING WATER

So where did all the water that must have existed earlier in Mars' history go? Some scientists have suggested that the northern plains may have once harboured large lakes or temporary oceans, and that the water still exists as ice deposits below the surface. Indeed the debris distribution pattern around impact craters provides indirect evidence for the existence of subsurface ice on Mars.

Many Martian impact craters are surrounded by ejecta which, at a casual glance, look like splashes of mud. Many astronomers believe these 'rampart craters' were formed by impacts that penetrated the subsurface ice or water, which then mixed with the material thrown out by the impact, causing it to flow along the surface. There are also peculiar polygonal patterns on the northern plains, which look like patterns seen on ice-rich ground in Arctic regions on Earth.

Further evidence for the existence of water on Mars was discovered in 1998, when Mars Global Surveyor detected high levels of hematite near the Martian equator. Such accumulations of this mineral typically grow in iron-rich water that has been heated by volcanic activity.

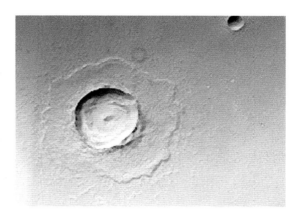

Belz Crater in Chryse Planitia, near the Martian equator. Belz is a typical example of what are called rampart craters, with a raised ridge or rampart around the inner layer of ejected debris. These craters may show evidence for the existence of subsurface ice on Mars.

The Parana Valles, a valley network in the Margaritifer Sinus region of Mars, photographed by the Viking I orbiter in 1976. These networks look similar to river drainage networks on the Earth and were presumably formed by running water sometime in Mars' past. The image reveals an area that is about 250km across.

The canal controversy 50 ▶

A Martian sunset over two prominent hills named Twin Peaks by Pathfinder scientists. The sky stays bright for up to two hours after sunset, indicating that dust particles extend very high into the Martian atmosphere.

PHOBOS AND DEIMOS

Mars is accompanied by two small moons, Phobos and Deimos, which are undoubtedly asteroids that were captured by the gravity of Mars. The small size of these moons means that, unlike the planets, they are not squashed into spheres under their own gravity, and are irregular potato-shaped bodies.

Phobos is 27km long and is dominated by a massive 10km diameter impact crater called Stickney. As well as being covered in many smaller craters, Phobos has a series of 100–200m wide grooves that may have been formed because of stress caused by the impact that created Stickney.

Deimos orbits farther out from Mars than Phobos, and is also extensively cratered, but smoother, darker and smaller than Phobos, measuring 11–15km across.

OBSERVERS OF MARS

Almost since the invention of the telescope in the early 17th century astronomers have been observing our nearest planetary neighbour. In the last quarter of the 17th century, Giovanni Cassini, Robert Hooke, Christiaan Huygens and others noted that Mars has bright polar caps that grow and shrink in a seasonal cycle. Huygens also deduced by observing the regularity with which permanent features came into view that Mars rotates in about one Earth day. In the late 18th century William Herschel noted the existence of temporary bright patches, which he, correctly, explained as clouds, and worked out that Mars' axis of rotation was tilted about 25° toward the plane of its orbit, which is why Mars, like the Earth, has seasons.

In 1837 Wilhelm Beer and Johann Mädler noted a dark band surrounding the shrinking north polar cap, and suggested that it might be a wet region. Astronomers now know that it was, in fact, caused by seasonal dust storms.

THE CANAL CONTROVERSY

In 1877 the Italian astronomer Giovanni Schiaparelli published his now-famous map of Mars. This map showed many light and dark features, including a network of dark linear markings that he described as *canali* (Italian for channels or grooves). The term was mistranslated into English as canal, and growing numbers of other observers reported the appearance of these dark lines. By the end of the 19th century canal mania began to sweep the astronomical community, with some convinced that the *canali* were the work of intelligent Martians.

Percival Lowell, a wealthy American with an interest in astronomy, was so entranced by the 'canals' that in 1894 he founded his own observatory in Flagstaff, Arizona, with the purpose of observing Mars. Lowell produced many charts, and championed the idea of canal-building Martians trying to irrigate their increasingly arid world. Schiaparelli was not convinced, favouring a natural origin for the *canali*, although he did not completely rule out Lowell's ideas.

As telescopes improved, a growing number of observers began to doubt the reality of Lowell's canals. Astronomers like Eugenois Antoniadi, and Evans and Maunder saw the canals as nothing more than optical illusions caused by poor observing conditions or inferior optics, or even the observer's brain trying to join up unconnected features with straight lines.

Photographs from the Mariner and Viking spacecraft showed that the canals do not exist, although many of the other dark markings shown on the telescopic maps do correspond to real geological forms on Mars.

Mars's satellites, Phobos (below) and Deimos (below right), are captured asteroids. They were discovered in 1877 by the American astronomer Asaph Hall.

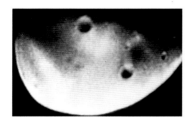

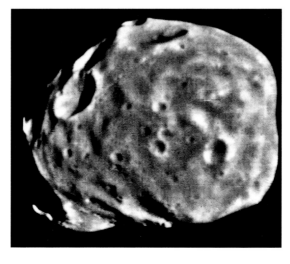

MARS' SATELLITES

Name	Interesting features
Phobos	A captured asteroid, about 27x22x19km, Phobos has two large impact craters, Hall and Stickney. Stickney is 11km in diameter. A network of pitted grooves about 100–200m wide and 20–30m deep radiate from Stickney. They are thought to be internal stress fractures resulting from the impact that formed the massive crater.
Deimos	Deimos is the smaller and farther out of Mars' two satellites, Another captured asteroid, it is about 15x12x11km. Its largest craters, such as Voltaire and Swift, are about 1.3km across.

EXPLORING MARS

A primary goal of the Viking landers was to search for life. Both Vikings carried a battery of experiments to test the Martian soil for signs of living matter. Soil samples were scooped brought inside the landers and tested to see if they released or exchanged any carbon – an indicator of life processes – or if they contained any organic (carbon-based) matter, without which life cannot develop.

Instead of life-like chemistry, two experiments turned up chemical reactions that seem to indicate the presence of highly reactive peroxides in the soil, while a third apparently showed more life-like reactions. A fourth revealed no organic matter in the soil.

Mars Pathfinder placed a small rover on Mars. Called Sojourner, this rover roamed around the surface taking close-up pictures and performing chemical analyses of the rocks that littered the landing site. Sojourner sampled a number of rock types including sedimentary and igneous (crystallized from a molten state).

The lander vehicle itself carried a stereoscopic camera on an extendible mast, which has returned 3-D colour images of the surface and, like Viking, has monitored the local weather at the landing site.

In September 1997, the Mars Global Surveyor probe entered orbit about the red planet to spend two years mapping the surface mineralogy, topography and gravity, and monitoring global weather and atmospheric processes.

MISSIONS TO MARS

Name	Date	Aims and achievements
Mariner 4	1965	Fly-by. Sent back images. Revealed that Mars is 'dead' and cratered.
Mariner 6 and 7	1969	Fly-bys. Examined atmosphere.
Mariner 9	1971	First man-made orbiter of another planet. Mapping.
Viking 1 and 2	1976	Orbiters: global mapping. Landers: first successful landings; biochemical experiments on the soil to see if life could exist; photographs of the Martian surface.
Phobos 2	1989	Orbiter: photographs of Martian moon Phobos.
Mars Pathfinder	1997	Lander and Sojourner rover. First 'bouncing' landing. Chemical analyses of different kinds of rocks.
Mars Global Surveyor	1997	Mineralogical mapping and weather monitoring. Discovery of Mars' weak magnetic field.

This panoramic view of the surface of Mars was taken by the Pathfinder imager on the lander in 1997. The Sojourner rover is next to the rock, Barnacle Bill. Also visible are all three petals of the lander, the front and back ramps and the deflated air bags.

The Sojourner rover of the Pathfinder mission examines the rock, Barnacle Bill. The rock slightly farther away and to the right is Yogi, and was also examined by Sojourner during the very successful Pathfinder mission in 1997.

Life on Mars? 52–53 ▶

Asteroids 80–81 ▶

Orbits 90 ▶

Robotic missions 148–149 ▶

Observing the planets 160–161 ▶

LIFE ON MARS?

When I first heard about the Martian meteorite I did not realize how great a part ALH 84001 would play in my life. I was part of a team who studied the meteorite, looking at carbon. The evidence we and other scientists found is not conclusive but is thought-provoking, indicating that there is a very real possibility that life did exist on Mars at one time.

BY DR MONICA GRADY

Above: ALH 84001, was found in the Allan Hills ice field in Antarctica in 1984. Scientists believe that it is a portion of a rock dislodged from Mars 16 million years ago, which fell to Earth about 13,000 years ago.

" I remember hearing, in February 1994, that a new Martian meteorite had been found in the US Antarctic meteorite collection. It was recovered in 1984 from the Allan Hills region, but only just recognized as Martian. The meteorite was large (about 2kg) and of a different composition from the other Martian meteorites already found, with abundant patches of bright orange carbonate minerals. This announcement was exciting for me as a petrologist, as are all announcements about unusual meteorites available for study, but I did not realize then just how much it would come to dominate my life!

The reason for ALH 84001's later rise to prominence was publication of images that suggested that ALH 84001 contained fossilized Martian bacteria. Much of the media did not seem to understand and pounced on the worm-like looking structures as pictures of the 'bacteria', and trumpeted messages such as 'Martian life discovered'. What was ignored was the minute size of the 'bacteria' and the scientific debate that raged around what these structures were. The publicity encompassed not only the scientists who made the claim, but President Clinton himself, and anyone studying the meteorite.

Dave McKay and his team have put forward a number of pieces of evidence to suggest the existence of a primitive fossilized Martian life-form within carbonates in ALH 84001. The unusual nature of the carbonates became the basis of their arguments: how they formed, their structure, the existence of organic carbon associated with them and the discovery of magnetite (iron oxide) grains within them.

Although it is generally accepted that the unusual carbonates were created on Mars, the way in which they were formed is still an active area of debate, and one that is crucial to whether or not they might contain 'biological' material.

There are three ideas of how the carbonates may have formed, only one of which proposes an environment that would allow life to survive: the carbonates were formed from a watery fluid at low temperature (below 100°C) in a location near the Martian surface. The other two theories propose an environment hostile to the hardiest of bacteria. Firstly, the carbonates may have come from a carbon dioxide-rich fluid deep within the Martian ground at high temperature (above 500°C), or secondly from shock melting when the meteorite was heated suddenly to high temperatures.

McKay and co-workers interpreted the shape of the magnetite grains within the carbonate as having a biological origin. However, some scientists argued against such an origin in terms of what the grains do not show: they are not linked together, as they would be if they had been produced within a living organism, and they are not accompanied by void spaces bounded by cell wall structures, as in fossilized bacteria from the Earth. At least three possible non-biological origins for the magnetite have been proposed: production from a gas at high temperatures, deposited from a fluid at low temperatures, or by decomposition of the carbonates themselves.

Since the mid-1980s, I have been part of a team of UK scientists investigating carbon in Martian meteorites to see whether there is any evidence for a carbon cycle on Mars. Organic matter (molecules of mainly carbon and hydrogen) is the basis of life. Its presence on Mars would indicate the possible existence of bacteria or plants at some point. We were invited to study specimens of ALH 84001 and found levels of organic carbon associated with the unusual carbonate that could not be dismissed simply as contaminants acquired after the meteorite landed on the Earth. Dave McKay's team cited our findings in their report.

Another argument put forward was the presence of magnetite grains in the carbonates. It was in fact the scanning electron microscope images of freshly broken surfaces of carbonate grains, showing parallel ridge-like features that

Although the present environment is hostile towards the survival of life, conditions in the past may have been very different.

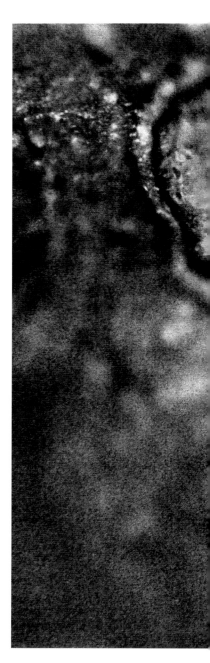

first sparked off the idea of the existence of fossilized bacteria in ALH 84001. The features, about 0.0002mm long, are too small for chemical analysis, but were inferred to be more magnetite grains. Criticism was initially levelled at the team who interpreted the features as bacterial, since they are much smaller than common bacteria. However, subsequent investigations have revealed that bacteria this small are much less uncommon than was thought at that time.

Concerns were raised that the features may have been produced during preparation of the samples, but these were allayed. Even so, doubts still arise as to whether it is valid to conclude that the features are bacteria just because they are bacteria-shaped.

ALH 84001 has had a complex geological history – it has endured at least two episodes of shock, has been heated and has absorbed fluids, all before it arrived on Earth to face 13,000 years in Antarctica. It then suffered the glare of global media publicity. Since August 1996, the number of papers (research and popular) written about ALH 84001 is probably sufficient to fuel an expedition to Mars to search for signs of life directly!

Currently, the balance in scientific thinking leans towards the belief that the evidence presented thus far is insufficient to conclude that life has been present on Mars. However, many scientists are agreed that, although the present environment is hostile towards the survival of life, conditions in the past may have been very different. It may have been much warmer and wetter, conditions under which biological processes might well have been viable. We won't know the truth about life on Mars, or otherwise, until we go to the Red Planet to search for it. **"**

The orange-coloured carbonate grains in ALH 84001 are believed to have been formed on Mars more than 3.6 billion years ago. How these carbonate grains formed is a major part of the debate on whether they provide evidence about the possibility that life once might have existed on Mars. This image shows an area of the meteorite that is approximately 0.5mm wide.

JUPITER

JUPITER IS THE GIANT PLANET OF OUR SOLAR SYSTEM, WITH A MASS MORE THAN TWICE THAT OF ALL THE OTHER PLANETS COMBINED.

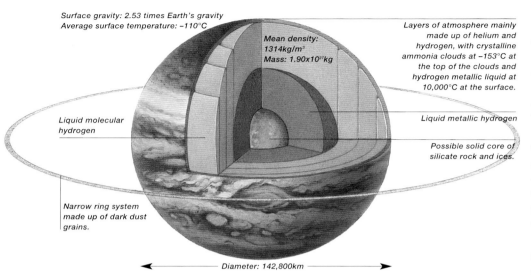

Surface gravity: 2.53 times Earth's gravity
Average surface temperature: −110°C

Mean density: 1314kg/m³
Mass: 1.90x10²⁷kg

Layers of atmosphere mainly made up of helium and hydrogen, with crystalline ammonia clouds at −153°C at the top of the clouds and hydrogen metallic liquid at 10,000°C at the surface.

Liquid molecular hydrogen

Liquid metallic hydrogen

Possible solid core of silicate rock and ices.

Narrow ring system made up of dark dust grains.

◄——— Diameter: 142,800km ———►

VITAL STATISTICS

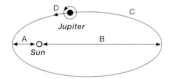

A: 740,900,000km
B: 815,700,000km
C: orbits in 4332.71 Earth days
D: rotates in 9hr 50min 30sec

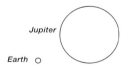

Jupiter

Earth ○

Jupiter's diameter is 11.19 times the size of the Earth's

ecliptic   equator
        orbit

Angle of equator to ecliptic: 3.13°
Angle of orbit to ecliptic: 1.31°

Satellites: 16

Jupiter, the fifth planet from the Sun, is striking not only because of its vast size but also because of the colourful bands of cloud that contain dramatic weather features, the largest of which is the Great Red Spot.

JUPITER'S BASIC CHARACTERISTICS
When viewing the planet, you can only see the outermost part of its very deep atmosphere, which has several layers of cloud of different composition and colour. Jupiter rotates so fast (a complete revolution in only 10 hours) that it spins the clouds into bands in which various spots, waves and other dynamic weather systems occur.

Jupiter has a striking, striped appearance, with white, yellow-brown and red clouds forming bands of various widths parallel to the equator. These have wavy patterns in them, especially along the edges, and giant spots can be seen between, and sometimes within, the bands. These are giant eddies, or rotating masses of cloudy air, similar to our earthly hurricanes. Other weather systems, often of contrasting colour, appear embedded in the cloud layers.

JUPITER'S WEATHER
The striking appearance of the banded patterns in Jupiter's clouds occurs because of the existence of convection cells in the atmosphere. Convection happens when a gas or liquid is heated. Imagine a pot of soup on a hob: the warmth from the hob heats the soup at the bottom of the pan, which then expands and rises. As it rises, the soup cools and sinks back to the bottom where it is heated again.

Jupiter is hotter at lower levels, which creates the convection cells in its atmosphere, causing moist gases to rise to levels where the volatile components (ammonia, water, etc) cool and condense to form the brighter regions seen at the cloud tops, called zones. The darker, mainly cloud-free belts are the regions where the relatively 'dry' air descends back toward the centre of the planet.

Because of the rapid rotation of Jupiter, the cells form narrow (compared to the size of the planet), symmetrical rings around it. The zonal winds along each belt at the cloud top level blow alternately eastward and westward, at speeds of up to 718km/hour relative to the deep interior.

ATMOSPHERIC LAYERS
Jupiter's atmosphere is made up of about 86 per cent hydrogen and 14 per cent helium by volume, and contains small amounts of methane (0.2 per cent), ammonia (0.02 per cent), water vapour (variable, but about 0.2 per cent) and numerous other compounds in proportions of less than one part per million.

The atmosphere is made up of layers, with varying temperatures and pressures. The coldest level in the atmosphere, the tropopause, has a temperature of about −153°C. Just below this level, ammonia condenses to form white, crystalline clouds, which are the deepest visible from Earth. The pressure at the cloud tops is about the same as that at which aeroplanes fly on Earth.

Lower down, at about the pressure of high mountain tops on Earth, lies another layer made of ammonia combined with hydrogen sulphide. Impurities in this cloud – possibly sulphur itself – give it a brown or yellow appearance.

Below that, at a pressure several times that at the Earth's surface, temperatures are above the freezing point of water. Scientists expect thick watery clouds to form at this depth. There are undoubtedly other types of clouds at even greater depths, which are not visible from the Earth.

At roughly 1000km below the level of the visible clouds, the pressure is high enough for hydrogen to change from the gas to the liquid form. The liquid hydrogen forms currents and convection patterns that carry heat from the inside to the outside of Jupiter.

Under the liquid hydrogen layer, at temperatures of around 10,000°C and at thousands of times the Earth's surface pressure, the hydrogen changes form again and becomes a metallic liquid, somewhat like mercury. At this great depth, the atoms are stripped of their negative electrons (ionized) and the remaining charged particles become metallic and conducting.

JUPITER'S MAGNETIC FIELD
Jupiter has a very strong magnetic field, which is caused by the fact that the planet is still cooling from its time of formation and is constantly collapsing in on itself under its own gravitational pull. This process gives off heat, which produces convective motions within the fluid metallic interior, making it act like a dynamo. This, coupled with the stirring action of Jupiter's rapid spin rate, is what produces Jupiter's enormously powerful and extensive magnetic field.

Jupiter's magnetosphere is enormous: Jupiter normally appears in our sky as a bright star, but if you were able to see the magnetosphere, it would appear as big as the Moon does in our sky. The Jovian magnetic field is about

20,000 times stronger than that of the Earth and it changes size and shape constantly in response to changes in the solar wind.

Overall, the magnetosphere is bullet-shaped, as opposed to the Earth's basically spherical field. This shape is caused because the magnetic field is not produced in the central core of the planet but by motions within the liquid metallic shell. Earth's field is more nearly a dipole, a magnetic field with two opposing poles. This is, in part, because it is produced nearer the centre of the globe. If we make the best fit we can of a dipole to Jupiter's field, then we find that it is inclined by about 10° to the rotation axis.

The magnetic field traps extensive bands of radiation around the planet and these, in turn, partially trap energetic charged particles of ionized atoms (from the solar wind and Io's volcanoes) and cause aurorae in the polar regions, similar to Earth's northern lights.

The satellite Io sits inside Jupiter's radiation belts. Since Io is volcanically active, it has an atmosphere that is continuously being removed by spluttering effects when the lava reacts with the particles in the radiation belts. The atoms and molecules that escape are ionized and end up in orbit around Jupiter, forming a plasma torus (a doughnut-shaped ring of ionized gas), which attracts the magnetic field of the planet and distorts it into a disk.

On the side of the planet away from the Sun, the field is stretched out by the solar wind into a long tail. This process is common to all planets, but in Jupiter's case the magnetotail is enormously long, stretching beyond the orbit of Saturn.

The largest planet in the solar system, Jupiter, is characterized by extensive cloud belts and giant eddies, such as the Great Red Spot. This image was taken by the Voyager spacecraft in 1979 and shows its satellite Io passing in front of it during its orbit.

The Great Red Spot 56 ▶

Io 57–58 ▶

Observing the planets 160–161 ▶

Jupiter's Great Red Spot varies in size but can measure 40,000 by 14,000km. Visible through small telescopes, it was first observed in the 17th century. The Great Red Spot is a huge eddy, colder than its surroundings. In this image, obtained by Voyager 1 in 1979, cloud details as small as 160km across can be seen below the spot.

THE GREAT RED SPOT

The largest eddy on Jupiter, the Great Red Spot (GRS), which occurs at a latitude of about 23° south, is a complex, cloudy vortex rotating in an anti-clockwise direction. It is larger than the diameter of the Earth and has existed since the time of the earliest recorded observations more than 300 years ago.

The GRS is believed to be an effectively two-dimensional vortex, which spirals outward away from areas of high pressure. This means that, although it looks like a giant version of a terrestrial hurricane, the GRS is a high- rather than a low-pressure feature and is much thinner compared to its width. The red-coloured clouds inside the GRS form vast, spiralling ribbons with narrow, relatively cloud-free gaps between them.

The latest results from the Galileo (1995) mission show that the outer edge of the GRS may be as much as 10km lower than the rest of the cloud within it. The outer part appears to rotate with a period of four to six days; near the centre, motions are small and more random in direction.

One theory suggests that such a very large, shallow disturbance may keep going by absorbing the angular momentum of smaller storms, when they run into it, rather than dissipating or breaking up which is the more familiar fate of weather systems on the Earth.

OTHER FEATURES

Other spots on Jupiter are smaller than the GRS but can still be very large; there are different types, mainly categorized by their shape and colour, for example white ovals. Nearly all of them rotate anti-clockwise, like the GRS, and last a long time: some white ovals visible now were first observed in 1939.

Low-pressure, cyclonic eddies do occur, characterized by their anti-clockwise rotation, although most of them do not last long, showing that they are relatively unstable on Jupiter, unlike on Earth.

Brown barges are a feature unique to Jupiter. Astronomers believe that they are holes in the clouds that let out the warmer radiation, which is generated in the interior of the planet. This is why they appear to be darker in colour than the white ammonia clouds surrounding them. They are unusual in that, although they are cyclonic disturbances, they are fairly long-lived.

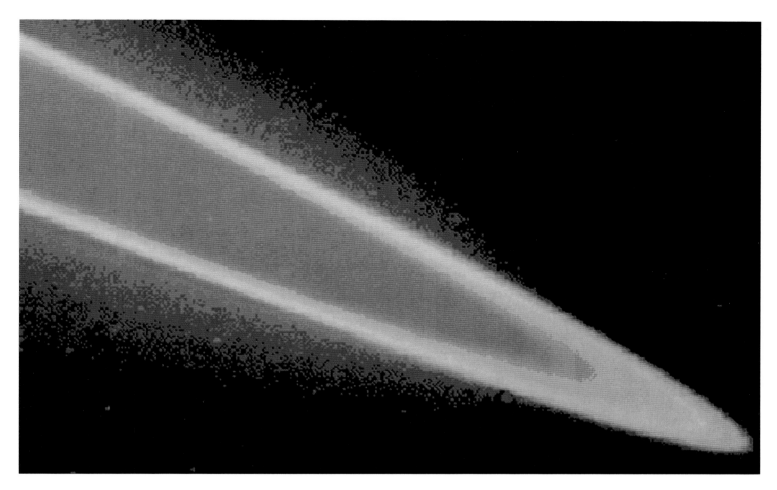

As would be expected from the rapid rotation and high winds, Jupiter's atmosphere shows lots of wave activity. This shows up in images of the planet: undulating edges of cloud bands, regularities in the positions of trains of whirlpools around the planet, and fluctuations in the size and shape of features and in the wind field, tracked by clouds. However, detailed understanding of Jovian meteorology is still in its infancy.

RINGS

Jupiter's ring system is composed of three major components. The main ring is 7000km wide and has an average radius of about 126,000km. At its inner edge the main ring merges gradually into the halo, a faint doughnut-shaped ring about 20,000km wide, which extends half of the distance to the visible disk of the planet. Just outside the main ring is a faint gossamer ring made up of very fine material, which extends out beyond the orbit of the satellite Amalthea.

The rings are not only much more tenuous than Saturn's, they are also darker, and probably made up of dust rather than ice. Dust grains have a fairly short lifetime in the powerful, fluctuating field and particle environments near Jupiter, and a source must exist to keep replenishing the rings. For the main ring this may be the two small moons, Adrastea and Metis, which orbit nearby plus, no doubt, other smaller bodies as yet undiscovered.

SATELLITES

There are 16 named moons that orbit around Jupiter, which is more than any other planet except Saturn and Uranus. Most of them are small and rocky and are more likely to be captured asteroids than original members of the Jovian system. The exceptions are the four largest, so-called 'Galilean' (after their discoverer) satellites, Io, Europa, Ganymede and Callisto.

The Galilean moons have diameters that, in units of the diameter of the Moon, are 1.04 (Io), 0.90 (Europa), 1.51 (Ganymede) and 1.38 (Callisto). Their physical properties are quite different from each other as a result of having evolved at varying distances from Jupiter.

IO

Io is about the same size as the Moon but more massive; at a mean density of 3530kg/m^3, it is the densest large object in the outer solar system.

Io is 421,600km away from Jupiter and is influenced so greatly by the giant planet's tidal effects that its surface is pulled upward and downward by hundreds of metres in each rotation. This flexing of the solid body produces some 10 megawatts of heating, sufficient to keep most of Io in a molten state.

Heat is also generated by perturbations from Io's sister satellite, Europa. The two moons are close enough and large enough to affect each other's orbits, with the net effect that Io (and, to a lesser extent, Europa) is constantly changing shape.

A thin ring system, composed of small rocky particles, around Jupiter was discovered by Voyager 1 in 1979. There are three named rings, the halo ring, main ring and the gossamer ring. This is a false colour, computer-generated image.

Europa 58–59 ►

Ganymede 59–60 ►

Callisto 60–61 ►

Saturn's rings 63–64 ►

Robotic missions 148–149 ►

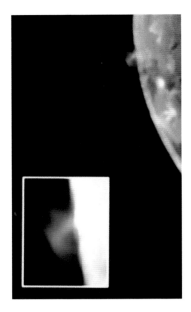

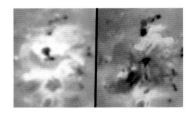

The Galileo spacecraft imaged a blue coloured volcanic plume on Jupiter's satellite Io in 1996. The plume extends about 100km above Io's surface and the blue colour is believed to be due to the presence of sulphur dioxide gas.

A comparison image of the volcano Ra Patera on Io showing the difference between the area imaged by Voyager in 1979 (left) and imaged by Galileo in 1996 (right). An area of about 40,000 square kilometres around the volcano has been covered by new volcanic deposits. These two images prove that Io is volcanically active.

Unlike Jupiter's other satellites, which are mostly made of ice, Io consists primarily of rocky material, such as silicates, making it more like the Moon or Mars. Furthermore, Io's interior is molten beneath its thin crust, as can be seen in the spectacular active volcanoes on its surface. The volcanoes mainly produce sulphur, though recent evidence suggests that the lava may, in fact, be similar in compostion to those on Earth. They produce a thin, transient atmosphere, probably composed mainly of sulphur dioxide. They also emit dust, made of sulphur compounds, which coats the surface and accounts for the orange, yellow and white colours. Lakes of hot sulphur lava many miles wide occur on the surface, while plumes of gas and particles shoot into the atmosphere to heights of 300km. The surface is constantly being replenished with lava and volcanic ejecta, which form a porous, low-density top layer.

The average daytime temperature on Io is around –150°C, but some areas where the crust is thin are at 30°C or so, and the temperature in the active volcanoes and lava lakes is above 330°C.

EUROPA

Europa is the smallest of the Galilean moons. Like Io, it is heated by the tidal effects of Jupiter and other satellites. But as Europa is farther from Jupiter the effect is not as great.

This image of Jupiter's moon, Io, was taken by the Galileo spacecraft in 1996, from a distance of 2.24 million kilometres. Io's surface is covered with volcanic deposits believed to contain silicate rock and sulphur-rich compounds, which give the satellite its distinctive colours. The dark areas are regions of current or recent volcanic activity. The brightest areas are sulphur dioxide frosts and the Media ridge area is visible in the middle of the image.

Astronomers expect that Europa's surface ice may be liquid at a depth of a few kilometres. The Galileo mission found evidence for cryovolcanism, where water ice is spurted out from the interior in a similar way to lava from volcanoes on Earth. This supposition is supported by other photographic evidence of shifting of blocks and plates of ice on the surface of Europa, and of the past outflow of liquid water from cracks in the ice in some regions.

Europa's surface is mostly frozen water ice. The most common features are different sorts of grooves and ridges, caused by compression of the crust or the emergence of warm ice from the interior. In addition, there are tilted and faulted blocks of terrain similar to sea ice seen on Earth and this supports the theory of an ocean on Europa's surface.

The shapes of Europa's surface features suggest that its orbit is not synchronous. The side of the satellite that trails as Europa orbits Jupiter is darker and redder than the leading side, probably as a result of asymmetrical bombardment by very small meteorites and by charged particles from Jupiter's magnetosphere.

Europa has the highest albedo of Jupiter's satellites, at around 65 per cent. This means that most of the sunlight is reflected and does not therefore reach the surface of the planet, leading to a maximum surface temperature of around −130°C; at the poles the temperature is probably around −220°C, not much warmer than on Pluto.

GANYMEDE

Ganymede is the largest satellite in the solar system with a diameter of 5262km, making it larger than Mercury and Pluto. It orbits Jupiter at a distance of just over one million kilometres and always keeps one face toward the planet.

At present, Ganymede is frozen more solidly than Europa because it is farther from Jupiter and less heated by

This image of Jupiter's satellite, Europa, was taken by Voyager 2 in 1979. Europa, about the size of the Moon, is thought to have a crust of water ice about 100km thick. The complex array of streaks indicates that the crust has been fractured and then filled with materials from the satellite's interior.

MISSIONS TO JUPITER

Name	Date	Aims and achievements
Pioneer 10	1973	First fly-by of Jupiter, 132,250km from cloud tops.
Pioneer 11	1974	Fly-by 42,900km from cloud tops.
Voyager 1	1979	Fly-by. Passed close to all four Galilean moons.
Voyager 2	1979	Fly-by and close to Callisto, Ganymede, Europa and Amalthea.
Ulysses	1992	Fly-by of north and south poles.
Galileo	1995	Orbiter and atmospheric entry probe. Measured atmospheric composition; made close images of Galilean satellites.

GALILEO'S MOONS
Jupiter was named by the Ancient Greeks and Romans after the king of the gods, the ruler of Olympus and patron of Rome. Not much more was known about Jupiter until 1610 when Galileo turned his telescope toward the giant planet. He saw four small points of light, which moved in the sky from night to night. He observed them over a period of time and deduced that they were orbiting Jupiter. He used this theory to explain that if these points of light orbited Jupiter, not everything in the universe went around the Earth, as was generally believed by his contemporaries. He pledged his support for the Copernican system, which was the first to state that each of the planets were orbiting around the Sun. The four points of light that Galileo saw around Jupiter are, of course, the Galilean satellites: Io, Europa, Ganymede and Callisto.

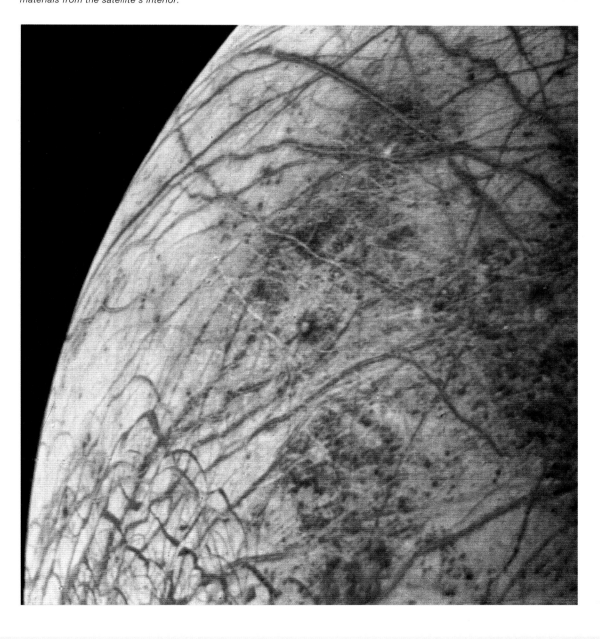

The internal structure of Pluto 76 ►

Robotic missions 148–149 ►

Ganymede, about one-and-a-half times the size of Earth's Moon, is composed of a mixture of rock and ice. The dark regions are ancient, heavily cratered terrain while the bright regions are younger craters surrounded by the rays of exposed ice. This image of Jupiter's largest satellite was taken by Voyager 1 in 1979.

This image is of the Uruk Sulcus region on Ganymede, taken by the Galileo spacecraft in 1996. The bright circular feature in the lower half of the image is an impact crater with some dark ejecta (material spat out by the impact) visible to the right.

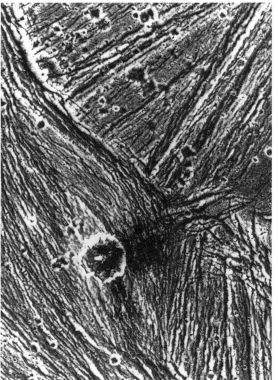

tidal effects. However, its surface and subsurface could still be somewhat malleable. The plate tectonics and groove formation may belong to an earlier era, when Ganymede was somehow subjected to stronger tidal coupling to Jupiter.

Its density is just less than twice that of water, implying a composition that is approximately half rock and half ice. Although Ganymede is largely covered by ice, it is much darker and dingy brown in colour over about half of its surface. The crust has broken up in places into dark 'continents' separated by fresher-looking icy material; astronomers think this is evidence for past tectonic activity.

The dark areas have a greater density of craters and are therefore probably older, perhaps by several billion years according to some interpretations of the cratering record, than the bright regions. The latter are younger areas, deformed by tectonic activity and characterized in many places by a complex of parallel grooves. The grooves occur singly and in bundles, and appear to date from the time the bright regions formed, perhaps by the upwelling of soft or even liquid material from below.

CALLISTO

Callisto is 4800km in diameter and 1,883,000km from Jupiter. As such, it is only weakly influenced by

Jovian tides and perhaps, unlike Ganymede, has always been that way. The crust would have frozen almost as it was formed, and little or nothing except for meteor impacts has happened on Callisto for 4.5 billion years.

Callisto, like Ganymede, is covered by a water ice crust, but there is no evidence for tectonic activity. Unlike the other Galilean satellites, Callisto is saturated with craters, suggesting very old terrain. Its surface is thought to have a coating of dark meteoritic dust. A large meteor produced the enormous crater, the Valhalla basin, which has a diameter of about 600km. The crater has ripples around it that formed when the surface ice was melted by the impact and then quickly refroze.

OTHER MOONS

The other 12 moons of Jupiter are much smaller than the Galilean satellites and range in size from 10 to 200km

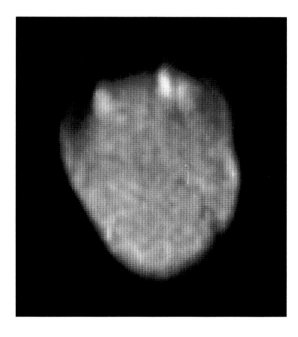

Jupiter's innermost satellite, Amalthea, orbits at a distance of just over one and a half times the size of Jupiter's radius in about 12 hours. Amalthea is an irregular body about 170x130km and is very heavily cratered.

in diameter. One of the innermost satellites is Amalthea, which orbits inside the orbit of Io and has a reddish colour, perhaps due to a coating of particles from the volcanoes on Io.

Beyond Callisto, at about 11 million kilometres from Jupiter, is a family of satellites consisting of Leda, Himalia, Lysithea and Elara. These all have orbits that cross each other and are inclined at about 28° to the planet's equator.

Another family of satellites, at about 23 million kilometres from Jupiter, contains Ananke, Carme, Pasiphae and Sinope, all of which orbit the planet in a backward direction. These unusual orbital properties tend to confirm that the small, rocky satellites are actually captured asteroids.

Callisto is a heavily cratered object with a number of ray systems. Its largest crater, the Valhalla Basin, is about 600km in diameter. This image of Callisto was taken by Voyager 2 in 1979.

JUPITER'S SATELLITES

Name	Interesting features
Metis	Small, irregularly shaped.
Adrastea	Small, irregularly shaped.
Amalthea	Its reddish surface may be a coating of particles from the volcanoes on Io.
Thebe	Small, irregularly shaped.
Io	The most volcanically active body in the solar system.
Europa	Covered in ice, possible with a liquid ocean beneath.
Ganymede	The largest moon in the solar system.
Callisto	Pock-marked with impact craters, indicating it is geologically dormant.
Ananke, Carme, Pasiphae and Sinope	All have retrograde orbits and are thought to be captured asteroids.
Leda, Himalia, Lysithea and Elara	Not much is known about these distant satellites.

Asteroids 80–81 ▶

SATURN

SATURN IS BEST KNOWN FOR HAVING THE MOST DRAMATIC RING STRUCTURE AND THE GREATEST NUMBER OF ORBITING MOONS IN THE SOLAR SYSTEM.

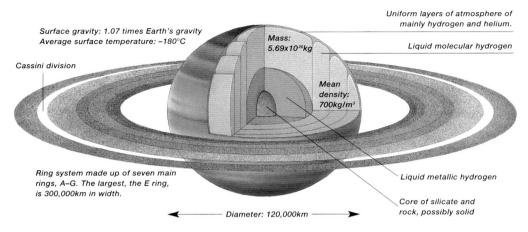

Surface gravity: 1.07 times Earth's gravity
Average surface temperature: −180°C

Uniform layers of atmosphere of mainly hydrogen and helium.

Liquid molecular hydrogen

Cassini division

Mass: 5.69x10²⁶kg

Mean density: 700kg/m³

Ring system made up of seven main rings, A–G. The largest, the E ring, is 300,000km in width.

Liquid metallic hydrogen

Core of silicate and rock, possibly solid

Diameter: 120,000km

VITAL STATISTICS

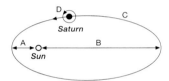

Saturn

Sun

A   B

C

D

A: 1,348,000,000km
B: 1,503,000,000km
C: orbits in 29.41 Earth years
D: rotates in 10hr 13min 59sec

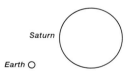

Saturn

Earth ○

Saturn's diameter is 9.26 times the size of the Earth's

ecliptic

equator

orbit

Angle of equator to ecliptic: 27°
Angle of orbit to ecliptic: 2.48°

Known satellites: 18

Saturn is striking because of its beautiful ring system. The planet is visibly flattened at the poles as a result of its very fast rotation on its axis and has the largest family of satellites in the solar system, numbering 18 in all.

SATURN'S ATMOSPHERE

The atmosphere of Saturn broadly resembles that of Jupiter. Saturn has 80–90 per cent hydrogen, 10–20 per cent helium and less than one per cent traces of other gases, including methane and ammonia, which have been detected by Earth-based and Voyager spectroscopy.

Because the cloud layers of Saturn are cooler than those of Jupiter, they tend to be thicker and more uniform in shape. They also form deeper in the atmosphere and the bands and spots – although present – are less pronounced than on Jupiter. Saturn's hazy yellow hue, and the deeper orange-yellow of one of its moons, Titan, are thought to be caused by deep haze layers of condensed hydrocarbons.

SATURN'S STRUCTURE AND COMPOSITION

Saturn is the second largest planet in the solar system: 1000 times the volume of the Earth, its density is less than that of water.

Current scientific models of the interior structure of Saturn suggest that the core of heavy elements makes up as much as a third of the mass of the planet. This implies a hot, rocky core with about three times the radius of the Earth, overlaid, as in Jupiter, by a liquid metallic hydrogen layer and a molecular hydrogen layer. It also seems that Saturn failed to retain as high a proportion as Jupiter of the light elements, such as hydrogen and helium, from its time of formation because its atmosphere is not as large.

Like its neighbours Jupiter and Uranus, Saturn radiates more energy into space than it receives from the Sun. The extra energy is generated by the gravitational contraction of the planet, whereby its surface is very slowly collapsing under its own gravitational pull. This heat production has been happening since the planet was formed, possibly augmented by the 'raining out' of helium deep in the interior.

Astronomers believe that temperatures in the core of Saturn exceed 12,000K, and the atmosphere must be deeply convective since this is the only plausible way to transport the interior heat to levels where it can be radiated to space.

Voyager 1 approached Saturn in 1980. Five days before closest approach, from a distance of 7.5 million kilometres, Voyager took this image of the structure of Saturn's atmosphere. This resembles Jupiter's atmosphere; although its features are less pronounced, bands, storms, ovals and eddies are all evident.

SATURN'S WEATHER

The convective atmosphere and rapid rotation of Saturn gives rise to cloudy bands like those on Jupiter. However, the markings are much fainter, with subtler colours than its giant neighbour, and the belts much wider, especially near the equator. This gives astronomers a clue as to what causes the belts and determines their pattern, but not enough is known for a comprehensive model of the two planet's atmospheres, explaining their belt structures, to be drawn up.

Wind speeds are very high at the cloud tops – near the equator, they reach velocities of 500m/sec (nearly 2000km/hr). This falls off at higher latitudes and alternates east and west in a pattern correlating with the cloud bands.

Saturn also exhibits some long-lived ovals and other features similar to those more common on Jupiter. In 1990 the Hubble Space Telescope observed an enormous white cloud near Saturn's equator caused by a planet-wide storm, which was not present during the earlier Voyager encounters; in 1994 another, smaller storm was observed.

MAGNETIC FIELD AND MAGNETOSPHERE

Saturn's magnetosphere is smaller than Jupiter's, but still extends well beyond the orbits of the outer moons. It is probably generated in the same manner as Jupiter's, by currents in the conducting layers near the core. However, the field is about 30 times weaker than that of Jupiter.

Saturn's magnetosphere is also simpler and more quiescent than Jupiter's, primarily because Saturn has no equivalent to the Io torus, the doughnut-shaped ring, which contains charged particles that react with the magnetic field. The magnetosphere, however, still produces aurorae and strong radio emissions, and reacts with the rings in its inner regions and with Titan in its outer regions.

RINGS

Saturn's ring system consists of hundreds, possibly thousands of narrow ringlets, which are grouped so that they give the impression of a small number of relatively broad rings. Each broad band has been given a letter.

The Hubble Space Telescope captured this image of the bright aurorae at Saturn's poles in ultraviolet light in 1994. The aurorae are produced when charged particles from the solar wind spiral down the magnetic field and enter Saturn's atmosphere at the poles. They then interact with elements in the atmosphere, producing these bright rings of activity.

SATURN'S RINGS

In legend, Saturn is portrayed as aged and lethargic, consistent with the slow apparent motion of the planet across the sky. Saturn's large ring system proved to be a challenge for astronomers. They were first discovered by Galileo through a small telescope in 1610. At first, he thought he had seen two large satellites, one either side of the planet; and it was only in 1655 that Christaan Huygens worked out that they were in fact a ring system. Giovanni Domenico Cassini (above), the first director of the Paris observatory, observed a gap in the rings, which now bears his name, and in 1785, Pierre Simon de Laplace proved that the rings could not be solid in structure. In 1857, James Clerk Maxwell proved theoretically that the rings are made up of independent solid particles.

◄ 16–19 Formation of the solar system

◄ 63 Saturn's magnetosphere

The bright A and B rings and the fainter C ring are visible through a telescope from Earth. The space between the A and B rings is called Cassini's division, while Encke's division, which is much narrower, splits the A ring. The Voyager camera detected four additional faint rings, named D, E, F and G, and two new divisions, the Maxwell and Keeler gaps, in the A and C rings.

This elaborate structure is due to the gravitational effects of the satellites, including some that are quite small, but which orbit close to and within the rings. One of these within the rings, now named Pan, was actually discovered as a result of a search for the cause of Encke's division. Another satellite called Mimas seems to have a corresponding role in creating Cassini's division.

Other so-called 'shepherding satellites' act to keep the rings in place by orbiting just inside and just outside of the ring material. They also set up waves in the rings which perturb them radially and vertically. The satellites Prometheus and Pandora shepherd the F ring, a complex structure made up of two narrow, braided, bright rings along which clumps of ring material, or possibly exceptionally large individual ring particles, are seen.

Voyager also photographed dark radial 'spokes' between Saturn and the wide B ring; these are probably caused by

About every 15 years, we see Saturn's rings edge-on because of the orbital geometry between Saturn and the Earth. The dark band visible above is the shadow of the rings. This is a Hubble Space Telescope image taken in 1995.

electrostatic charging of the ring particles and interactions with the planet's magnetic field.

Saturn's rings, as well as being much more massive than those of the other giant planets, are also much brighter, with albedos in the range 0.2–0.6. The rings are no more than 1.5km thick and are made up of thousands of small particles ranging in size from 1cm or so to several metres. They are composed at least in part of water ice, possibly with some rocky particles with icy coatings.

The origin of planetary rings is unknown. One theory is that they were formed out of the debris from the break-up of a satellite. If this is the case the satellite would only ever have been about 100km in diameter. The rings are not expected to last for the entire lifetime of the solar system, and were formed more recently than the planet, or have an unknown source from which they are regenerated.

SATELLITES

Saturn has 18 named satellites. Six of Saturn's moons are icy, and resemble the three outer Galilean satellites of

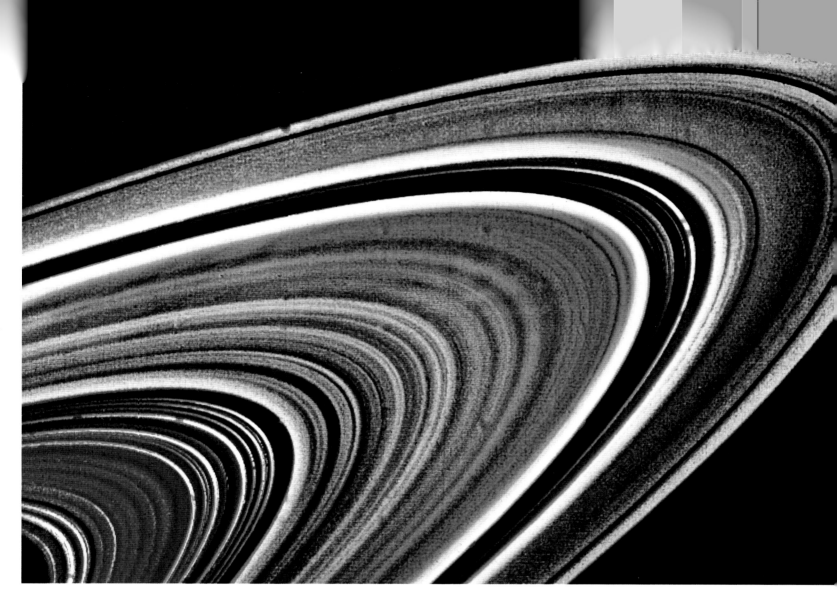

Jupiter; the others are small and rocky. Titan, with its thick atmosphere, is the odd one out.

The icy satellites – Mimas, Enceladus, Tethys, Dione, Rhea and Iapetus – all have densities less than 2000kg/m³, which implies that they are composed of 30–40 per cent rock and 60–70 per cent water ice. All of the icy satellites reflect at least 60 per cent of the light that strikes them, but Enceladus is an exception in that it

reflects almost 100 per cent, suggesting that it has a surface made up of relatively clean ice.

The surfaces of Enceladus, Rhea, Dione and Tethys all appear to be very similar; all are lightly cratered, with wispy white streaks across the surface. The two sides of Iapetus, however, contrast greatly in brightness; the side that leads in its orbit is very dark with an albedo of only 0.05, while the trailing side is 10 times brighter. This implies that the

Detailed images of the rings of Saturn show that they are made up of many thin ringlets. This enhanced colour view from Voyager 2 in 1981 shows possible variations in chemical composition.

Far left: This Voyager 2 image shows the satellite Prometheus and part of the F ring against the background of Saturn. Prometheus orbits Saturn at a distance of 139,350km and is about 148x100x68km. Prometheus may act as a 'shepherd' to the F ring. Its brightness suggests that it is an icy body like the larger satellites and ring particles.

Left: The huge crater, Herschel, on Saturn's moon Mimas is about one-third of the size of the satellite with a diameter of 130km. It is about 10km deep and has a central peak 6km high.

Titan 66–67 ▶

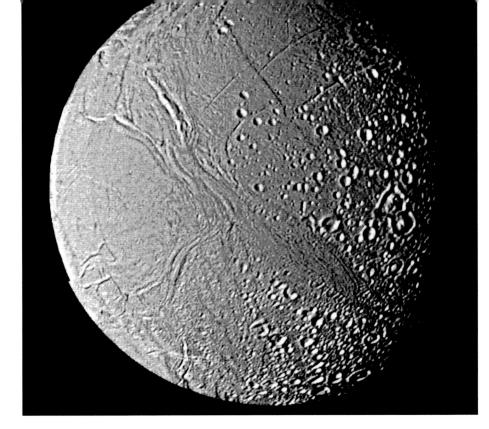

A large part of the surface of Enceladus is craterless, which indicates recent geological activity has wiped out older formations. The diagonal feature running from top left to bottom right is thought to be a huge ice flow. This image was taken by Voyager 2 in 1981.

satellite has swept up some black dust at some time in its history, or that very small meteors have eroded the bright surface away, exposing dark material underneath.

The satellites increase in size the farther they are from Saturn; the smallest, Mimas, is less than 200km in radius and is the smallest satellite known that has acquired a near-spherical shape. It is heavily cratered, and has one huge impact crater. The collision that caused this must have come very close to shattering the satellite.

Most of the satellites have nearly circular, synchronous orbits (meaning the same face is always directed towards Saturn) and lie in the equatorial plane, the exceptions being the outermost three. Distant Phoebe moves in a backwards orbit, which is inclined to the equator of Saturn, suggesting that it was probably an asteroid captured by the gravitational pull of the planet. Hyperion has an unusual cylindrical shape and tumbles chaotically in its eccentric orbit around Saturn because of gravitational interactions with Titan. Epimetheus and Janus orbit together about 50km apart and may be pieces of what was once a single satellite.

TITAN

Titan is the second largest satellite in the solar system, after Jupiter's Ganymede, and the only one among them with a thick atmosphere. This atmosphere is roughly 90 per cent nitrogen and the remaining 10 per cent other molecules such as methane. It is the only other place in the solar system with an atmosphere made up of the same major constituent as the Earth's.

The reason Titan has this thick atmosphere, unlike any other moon, is its distance from the Sun. The protoplanetary nebula, the cloud from which the planets were originally formed, was colder near Saturn than Jupiter and cold enough to allow nitrogen to condense; temperatures were too warm near Jupiter for this to occur on Ganymede. In the early phase of formation, Titan and Triton (Neptune's largest moon) probably both contained nitrogen ice, but now Titan is warm enough for nitrogen to evaporate to form an atmosphere, while the nitrogen on Triton is still frozen.

MISSIONS TO SATURN

Name	Date	Aims and achievements
Pioneer 11	1979	Fly-by; closest approach of 20,900km above cloud tops. Passed through ring plane twice. Returned 440 images.
Voyager 1	1980	Fly-by; approached within 125,000km and near Titan.
Voyager 2	1981	Fly-by within 101,300km, close to several satellites.
Cassini	2004	Orbiter and atmospheric Huygens entry probe.

Tethys' surface is completely covered by craters. The largest crater is called Odysseus and there is also a huge valley, Ithaca Chasma, more than 2000km, which runs three-quarters of the way around the satellite.

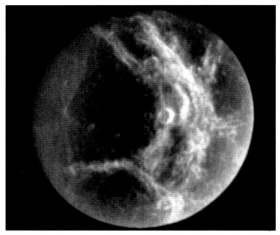

Dione displays a variety of features including faults, valleys, bright spots and depressions. There are also fractures that seem to be linked to the outgassing of water and perhaps methane.

SATURN'S SATELLITES

Name	Interesting features
Pan	Icy body 10km in radius.
Atlas	Shepherd satellite for A ring.
Prometheus and Pandora	Act together as shepherds for Saturn's F ring.
Epimetheus and Janus	These satellites orbit about 50km apart, and may be pieces of what was once a single satellite.
Mimas	Has a crater one-third of its diameter on it.
Enceladus	Has a strange grooved terrain superimposed over an older cratered surface.
Tethys	Largest of inner satellites, heavily cratered and probably made up mainly of water ice.
Telesto Calypso	Co-orbital satellites of Tethys.
Dione	Has some craters measuring over 150km in diameter.
Helene	Co-orbital satellite of Dione.
Rhea	Second largest of Saturn's satellites.
Titan	The second largest moon in the solar system. Has a thick, cloudy atmosphere.
Hyperion	An elongated satellite with a chaotic spin.
Iapetus	Its leading hemisphere is much darker than its trailing hemisphere, which is mainly covered in water ice.
Phoebe	Has a retrograde orbit and is probably a captured asteroid.

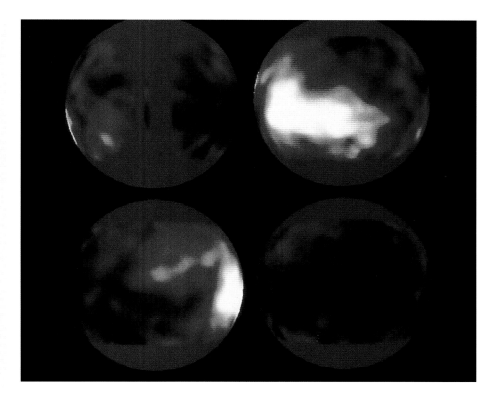

The temperature in Titan's lower atmosphere ranges from about –200°C near the surface to about –180°C at a slightly higher level. This makes it warmer than Saturn's airless satellites because of the greenhouse effect, but cold enough for methane and ethane to condense. Some models, therefore, predict cloud layers of both constituents, overlaid by a deep haze of oily liquid droplets, produced by the action of sunlight on atmospheric methane. These materials may rain out and form lakes on or under the surface.

The outer part of Titan's solid body is primarily composed of water ice in different phases (possibly including a liquid water 'ocean' at some depth below the surface), overlying a core of heavier rocky and metal elements such as silicates and iron, similar to the other icy moons.

This series of Hubble Space Telescope images of Saturn's cloud-shrouded moon, Titan, reveal light and dark features. The prominent bright area is a surface feature that scientists have not yet identified.

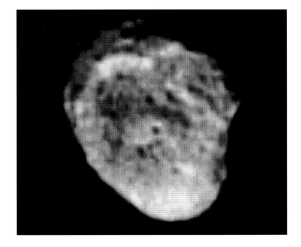

Hyperion is as large as Mimas but has a distinctly irregular shape, which may have been caused by a past collision. It is made up primarily of ice with some rocky impurities.

Iapetus has one bright and one dark hemisphere. A patch of dark terrain is visible to the top right of this image, the origin of which is still not fully understood.

URANUS

TILTED ON ITS AXIS, URANUS MOVES AROUND THE SUN ON ITS SIDE AND SUFFERS THE MOST EXTREME SEASONAL CHANGES OF ANY PLANET IN THE SOLAR SYSTEM.

Surface gravity: 0.92 times Earth's gravity
Average surface temperature: –216°C

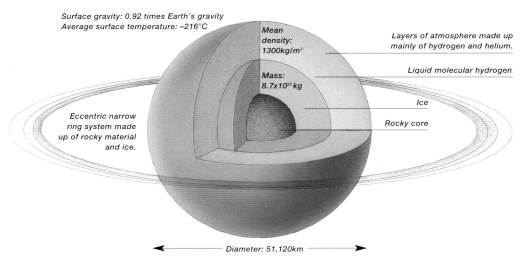

Mean density: 1300kg/m³

Mass: 8.7x10²⁵ kg

Layers of atmosphere made up mainly of hydrogen and helium.

Liquid molecular hydrogen

Ice

Rocky core

Eccentric narrow ring system made up of rocky material and ice.

← Diameter: 51,120km →

VITAL STATISTICS

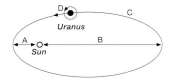

A: 2,739,000,000km
B: 3,003,000,000km
C: orbits in 84.04 Earth years
D: rotates in 17.2 hr

Uranus

Earth

Uranus' diameter is 4.01 times the size of the Earth's

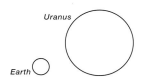

ecliptic
orbit
equator

Angle of equator to ecliptic: 97.9°
Angle of orbit to ecliptic: 0.77°

Satellites: 17

A smooth aqua-coloured sphere with very subtle hints of bands – that is how Uranus might appear to someone looking at it from one of its moons. The planet's calm facade gives no hint of a history fraught with spectacular catastrophe: at some stage in Uranus' past, a mighty collision wrenched the young planet off its axis. As a result, the planet is tipped over on its side so its rotation axis lies almost in the plane of the planet's orbit, giving rise to the most striking seasonal changes. Another collision may have been responsible for the fantastic geology of its moon Miranda.

URANUS' ATMOSPHERE

The basic composition of Uranus is the same as the other giant planets and similar to that of the Sun: it is predominantly hydrogen (about 80 per cent) with some helium (15 per cent). The remainder of Uranus' atmosphere is methane, hydrocarbons (molecular mixtures of carbon, nitrogen, hydrogen and oxygen) and other trace elements. Uranus' colour is caused by the small amount of methane – less than three per cent – that preferentially absorbs red light. So the reflected sunlight that we see is greenish-blue.

URANUS' COMPOSITION

Like Jupiter and Saturn, Uranus is a gas planet. We see only the outermost layers of clouds, which are composed of icy droplets of condensed methane and other materials like ethane and acetylene. The temperatures at this atmospheric level are very cold – about –200°C. Below this outer layer of clouds, the atmosphere gets thicker and warmer.

As the pressures and temperatures rise with depth, the hydrogen and helium transform from a gaseous state to a liquid state. At greater depth, a transition occurs to a thick, viscous, partly solidified layer of highly compressed liquid water, which may have traces of ammonia and methane.

Deep within the centre of Uranus at extremely high pressure, a core of rocky material is hypothesized to exist, with a mass almost five times that of Earth.

URANUS' MAGNETIC FIELD

When the Voyager 2 (1986) spacecraft flew by Uranus, it detected a magnetic field. The strength, about 50 times stronger than that of the Earth, was not unusual. What did surprise scientists was the apparent source of the magnetic field – it seemed to be centred on the outer edge of the rocky core as opposed to in the centre of the core itself. Furthermore, the magnetosphere was tilted nearly 60° from the planet's rotation axis. Because it was offset and tilted, the magnetic field wobbled and tangled as the planet rotated.

The magnetic field is responsible for auroral activity; Voyager 2 detected aurorae that seemed to be excited by very energetic electrons. Usually, aurorae are caused by charged particles from the solar wind, which spiral down the magnetic field and react with the atmospheric gases. However, in an enduring puzzle, there does not seem to be an adequate source for these electrons around Uranus.

From variations in the magnetic field strength, Voyager 2 determined that the planet's internal rotation period, the rotation of the inner rocky core, was 17.2 hours. The winds in the cloud layers have rotation periods ranging from about 16–18 hours depending on latitude, implying wind speeds reaching 200 metres per second for some regions.

URANUS' LACK OF HEAT

One of the most puzzling aspects of Uranus is the lack of excess heat radiating from its interior. Although the interior is warm, that warmth can be accounted for as part of the natural equilibrium of the atmosphere. In contrast, the other three giant planets radiate much more heat than expected from equilibrium conditions; this is believed to be residual heat from the time of the planets' formation and from continuing gravitational contraction.

Scientists theorize that perhaps Uranus also has this heat but that it is trapped by atmospheric layers, or perhaps the event that knocked Uranus on its side somehow caused much of the heat to be released early in its history.

URANUS' RINGS

In 1977, astronomers were watching as Uranus passed in front of a faint star (this is called an occultation). By measuring the star's light as it disappeared into Uranus' atmosphere, they hoped to deduce the temperature, pressure and composition of the outermost cloud layers.

Before the occultation began, they had their instruments running, and were very surprised when the starlight winked off and on several times as Uranus approached. After the occultation was over and Uranus was moving away from the star, the starlight again flashed off and on a number of times. They had discovered that Uranus has a ring system.

Uranus has nine well-defined rings, plus a fainter ring and a wider fuzzy ring. Unlike the broad system of Saturnian rings, the main Uranian rings are narrow. The thinness of the rings has led to the speculation that the gravity of small moons is keeping the rings confined.

The rings are not perfectly circular. The epsilon ring in particular is very eccentric and varies considerably in width (from 20km up to nearly 100km).

Like the rings of Saturn, the Uranian rings are thought to be composed mainly of rocky material (ranging in size from dust particles up to house-sized boulders) mixed with small amounts of ice.

Since their discovery, the rings have been monitored over a number of occultations, and were directly imaged by the Voyager 2 spacecraft. By looking back toward the Sun from the far side of the rings, Voyager 2 was able to see dusty material between the narrow rings, though there was much less dust than expected based on the Saturnian system. With new infrared cameras on the ground and with the Hubble Space Telescope, we can now see the rings directly from Earth.

MISSIONS TO URANUS

Name	Date	Aims and achievements
Voyager 2	1986	Fly-by discovered many small satellites, including Juliet, Portia and Bianca.

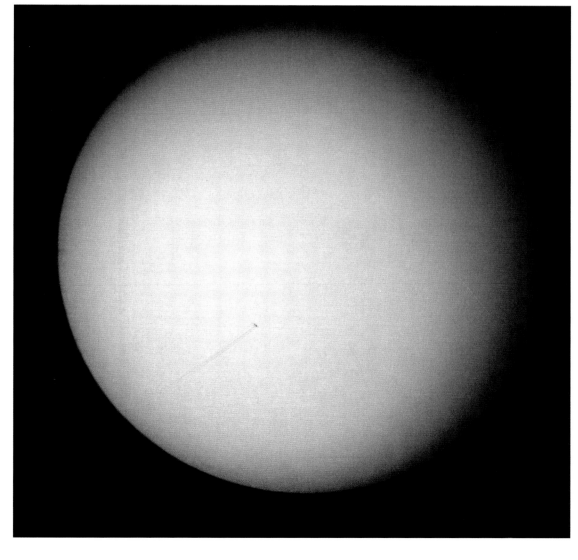

Above: Uranus' rings are made up of tiny dark particles. This image, taken by Voyager 2 in 1986, shows part of the ring system. The short bright streaks are trails of background stars.

Left: A true colour image of Uranus taken in 1986, by Voyager 2. Uranus appears blue because the methane in its atmosphere absorbs red light.

Right: This image of Oberon taken in 1986 by Voyager 2 shows some large craters with bright rays emanating from them and dark patches on their floor.

Far right: This 1986 Voyager 2 picture of Titania shows its icy surface with many small craters, a few large impact basins and an extensive network of faults up to 5km deep.

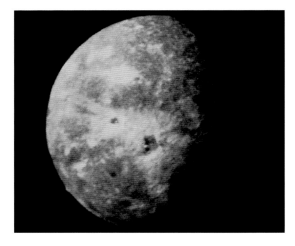

HISTORY OF DISCOVERY
Uranus was the first planet to be discovered that had not been known since antiquity. Although it is just bright enough to be seen with the naked eye, and in fact had appeared in some early star charts as an unidentified star, William Herschel was the first to recognize it as a planet in 1781. William Herschel made his own telescopes, including grinding his own mirrors. He was unable to name the planet after the then king of England, George III, but his discovery earned him a royal pension to continue his studies.

MYRIAD MOONS

Within six years of the discovery of Uranus, two moons were discovered circling the planet and were subsequently named Titania and Oberon. For more than 60 years thereafter, astronomers searched for more moons without success. It was only after the discovery of Neptune that two more Uranian moons, Ariel and Umbriel, were found in 1851. Nearly a century elapsed before Miranda was discovered in 1948, bringing the total of large moons to five. In 1986, Voyager 2 imaging revealed 10 minor moons, with the moon Puck being the largest.

In 1997, astronomers using the Palomar 5m telescope discovered two new moons temporarily named S/1997 U1 and S/1997 U2. These small moons follow eccentric and inclined orbits. Although this brings the total number of Uranian moons up to 17, astronomers speculate that there are more moons still to be discovered.

The moons of Uranus are presumed to be primarily rocky with some component of ice; the largest moons are big enough that the interiors had sufficient pressure and thereby heat to melt and cause tectonic activity called cryovolcanism (with icy flows rather than hot lava flows). However, the moons are all relatively small, grey and very dark, reflecting little sunlight. Thus, little was known about their surface structure or history until the Voyager 2 spacecraft returned detailed images of the surfaces of the main moons.

MIRANDA

Miranda, living up to its name, provided the most remarkable images. Huge geological features dominate the small moon's landscape, indicating that some kind of intense heating must have taken place at some stage in its past. Ridges, grooves, cracks and cliffs crowd together to form a wrinkled surface. It is not yet clear whether a massive collision disturbed the small satellite, which then reassembled into the current jumble, or whether, in the past, tidal interactions with other satellites produced the heat to melt and modify the surface, as is the case for Jupiter's moon Io.

TITANIA AND OBERON

Titania and Oberon are the largest Uranian moons and, like Umbriel and Ariel, form a pair with similar size and density. Oberon, the outermost major moon, shows a number of craters with brightly coloured ejecta around them. It has many large craters, and a number of these craters seem to contain old volcanic flows on their floors.

Titania has fewer large craters, indicating that its surface has been 'wiped clean' by volcanic resurfacing sometime in the moon's past. Titania's surface is also riddled with fractures, further evidence for extensive volcanic activity.

ARIEL AND UMBRIEL

Although the other large moons have less visible geologic activity, their surfaces are still diverse. Ariel has the youngest surface of the major moons, based on cratering rates. Evidence for past global tectonic activity includes fractures, linear grooves and smooth patches which may be the result of glacial-like flow of material.

Umbriel, Ariel's 'twin' in size and density, is much darker and smoother indicating that, unlike Ariel, it did not undergo tectonic activity. Its heavily cratered surface is probably the oldest of the satellites.

TINY PUCK AND THE MINOR MOONS

Voyager 2 was able to return nearly 50 images of tiny Puck, showing it to be an irregularly shaped body with a mottled surface.

The other nine minor moons discovered by Voyager 2 were seen in 38 different images, but the spacecraft did not get close enough to resolve their shapes or surfaces. Little is known about the newest moons other than their sizes and orbits.

STUNNING SEASONS

Uranus has often been called featureless, bland and even boring. These epithets are a consequence of fate and unfortunate timing. It was fate that caused a catastrophic collision of Uranus with a large body sometime early in the planet's development. This event was so energetic that the bulk of the planet, which was originally spinning anti-clockwise in the plane of the solar system as the other planets do, was bent over by an angle of more than 90°. (This gives rise to the curious circumstance that the north pole of the planet is now located just below the plane of the ecliptic, that is, in the south.) The orbits of Uranus' moons and its ring system are also tilted at this same extreme angle.

Left: Voyager 2 took many pictures of Ariel. This image taken in 1986 shows many craters and broad, branching, smooth valleys.

Far left: This false coloured Voyager 2 image of Umbriel taken in 1986 shows its heavily cratered surface. Its most prominent feature is an impact crater, 100km in diameter with a bright central peak, which is visible in the top right-hand corner of this image.

URANUS' SATELLITES

Name	Interesting features
Puck	Irregularly shaped with a mottled surface.
Miranda	The surface seems to show evidence of repeated phases of shattering and reformation.
Ariel	Has the youngest surface of the major moons.
Umbriel	Its heavily cratered surface is old compared to other large moons
Titania	Has prominent ice cliffs and fault lines, so it may be geologically active.
Oberon	Heavily cratered, with apparent 'volcanic' flows on some large crater floors.
Cordelia and Ophelia	Small satellites that 'shepherd' Uranus outermost ring.
Juliet, Portia, Bianca, Cressida, Desdemona, Rosalind and Belinda	Not much is known about these satellites except for their sizes. They were discovered by Voyager 2 in 1986.
S/1997 U1 and S/1997 U2	Discovered in 1997, these are probably captured asteroids.

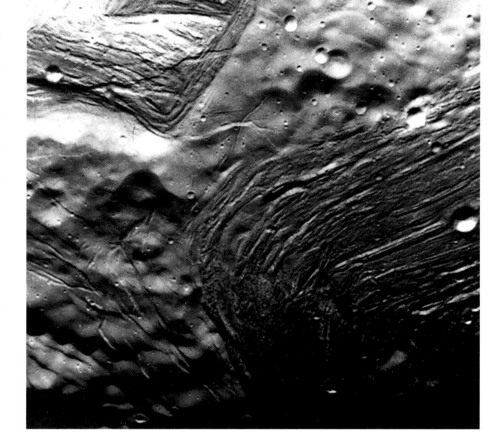

A consequence of Uranus' large axial tilt is that the planet's seasons are extreme: during the southern solstice, the south pole bakes in sunlight for nearly 20 years, with the northern pole losing heat into freezing blackness. Only during times of equinox, when we see the planet from the side, does the incoming sunlight fall on Uranus like it does on the other giant planets. Records show that during times of solstice, Uranus appears bland and featureless. Reports at times of equinoxes indicated clouds and banded structure.

BAD TIMING

It was unfortunate timing that the Voyager 2 encounter occurred at peak southern solstice. Its best quality images record only the subtlest of bands and the smallest smattering of clouds. Uranus is undergoing a face-lift at present. As it continues its 84-year-long progression around the Sun, its equatorial region is now receiving regular sunlight again, and parts of its northern hemisphere are being bathed in solar radiation for the first time in decades.

Today, images from the Hubble Space Telescope are revealing multiple bright cloud features and stunning banded structure on Uranus. Thus far, these features are most visible in the near infrared, but some are even detectable at visible wavelengths. It is fascinating to speculate how Uranus will appear to us by the time it reaches equinox in 2007. One thing we can be sure of is that Uranus will no longer be called boring.

This close up view taken in 1986 by Voyager 2 of the surface of Miranda shows light and dark grooves with their sharp boundaries. The area shown is about 150km in diameter and details as small as 600m wide can be seen.

The quest for aperture 136 ►

The electromagnetic spectrum 140–141 ►

Hubble's impact 144–145 ►

Star maps 172–181 ►

NEPTUNE

NEPTUNE IS A DYNAMIC SPINNING BLUE BALL OF
GAS, WHICH IS SITUATED IN THE FAR REACHES
OF OUR SOLAR SYSTEM.

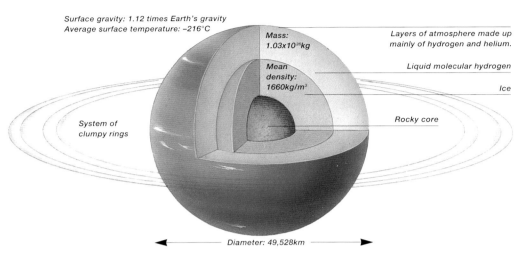

Surface gravity: 1.12 times Earth's gravity
Average surface temperature: –216°C

Mass:
1.03x10²⁶kg

Mean
density:
1660kg/m³

*Layers of atmosphere made up
mainly of hydrogen and helium.*

Liquid molecular hydrogen

Ice

Rocky core

*System of
clumpy rings*

◄——— *Diameter: 49,528km* ———►

VITAL STATISTICS

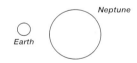

A: 4,456,000,000km
B: 4,546,000,000km
C: orbits in 164.8 Earth years
D: rotates in 16.11 Earth hr

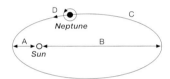

*Neptune's diameter is 3.88 times
the size of the Earth's*

Angle of equator to ecliptic: 1.77°
Angle of orbit to ecliptic: 29.6°

Satellites: eight

Until the Voyager 2 mission of 1989, we knew very little about this distant blue planet. We have now learnt that it is a huge gassy planet surrounded by clumpy rings and eight moons. It has an incredibly dynamic atmosphere that is constantly evolving and changing, which means that some features observed by the Voyager 2 spacecraft have now mysteriously disappeared. This faraway planet still possesses a great many unsolved secrets for us to unravel.

THE FINAL GIANT

Neptune is the last of the giant planets. A near twin to Uranus in size, Neptune has a similar atmospheric composition (about 80 per cent hydrogen, 15 per cent helium, 3 per cent methane, and other trace elements) and internal structure (a rocky core surrounded by a watery mantle that is rich in methane and ammonia and topped with a thick atmosphere).

Like the Uranian magnetic field, Neptune's field is also offset from the planet's centre and significantly tilted with respect to the planet's rotation axis. This indicates that whatever process creates such fields is not unique to Uranus. Current theories postulate that a thin electrically conducting shell around the planet's core powers the offset, tilted field. Neptune's field is about 60 per cent weaker than that of Uranus.

There are, nevertheless, important differences between Neptune and Uranus. The icy particles in the uppermost cloud decks of Neptune are of slightly different composition than those of Uranus; their bluish colour, combined with the atmospheric methane that absorbs red light, gives Neptune a richer blue tint compared with more greenish Uranus.

Unlike Uranus, which has no detectable internal heat source, Neptune has the strongest internal heat source of all the giant planets. It radiates almost three times more heat than equilibrium conditions would predict, as opposed to Jupiter and Saturn, which radiate about twice as much energy than expected. Neptune is thought to have excess heat from the time of its formation, which was a slower process than that of the other planets and from continued gravitational contraction, whereby the surface of the planet keeps collapsing under its own gravitational pull. Other as yet unidentified sources may also contribute heat.

NEPTUNE'S DYNAMIC ATMOSPHERE

Clouds and storms are the main features of Neptune's dynamic atmosphere, though the planet's distance from Earth makes it very difficult to observe them. Before the Voyager 2 (1989) encounter we knew that bright and highly variable clouds existed. Even so, the diversity of Neptune's atmospheric structure revealed by Voyager stunned astronomers.

Dominating all was the Great Dark Spot, a hurricane-like storm in the southern hemisphere, which was about half the size of the Earth. The next feature to be discovered was a small white spot, which appeared to race rapidly around the planet when compared with the stately Great Dark Spot. It was named the 'Scooter'. Many more spots were detected, some rotating even faster than the Scooter.

One of the most stunning atmospheric images was taken near the edge of the planet: several streaks of white high-altitude clouds were seen casting shadows on a deeper cloud deck far below them. This was the first direct detection of layers in an outer planet atmosphere.

All Neptune's weather conditions are developing and evolving rapidly. The Great Dark Spot changed shape with every image. Wispy bright clouds appeared and dissipated with each revolution of the planet. A small dark spot, D2, developed a bright core, while a bright clump near the south pole was composed of fast-moving bright patches.

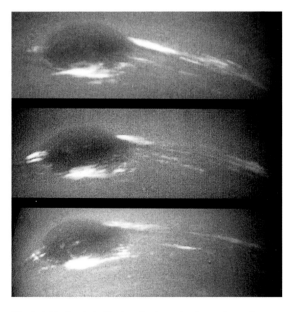

The bright clouds in Neptune's dynamic atmosphere change rapidly, often forming and dissipating over several hours. In this 1989 sequence spanning two rotations of Neptune (36 hours), Voyager 2 observed cloud evolution in the region around the Great Dark Spot. The surprisingly rapid changes, which occurred over the 18 hours separating each panel, show that in this region Neptune's weather is perhaps as dynamic and variable as the Earth's. However, the scale is immense by our standards because the Earth is only the size of the Great Dark Spot.

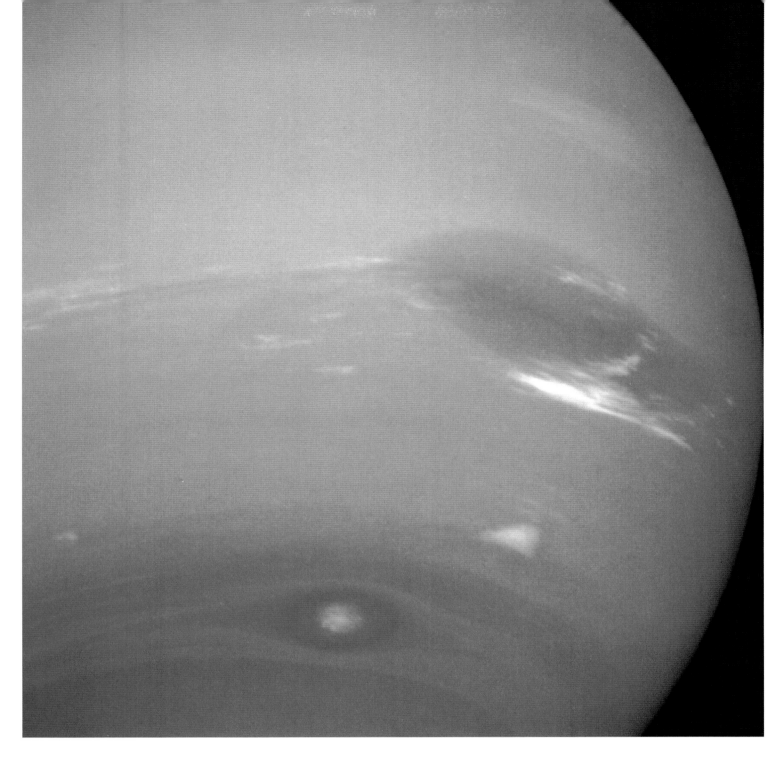

HIGH WINDS

Voyager measured the 16.11-hour internal rotation period (the rotation of the rocky core of the gas giants) by monitoring the magnetic field. However, the rapid evolution of the cloud features made it hard to track atmospheric rotation periods.

Careful study showed that the atmosphere rotated with periods ranging from over 18 hours near the equators to faster than 13 hours near the poles. In fact, the winds of Neptune are among the fastest in the solar system, at 700m/sec, dwarfed only by Saturn's high-speed equatorial jet.

The cause of Neptune's dynamic weather might be the tilt of its rotational axis. At only 29°, it is more like Jupiter than Uranus.

TRITON

Neptune's largest moon is Triton. Because of its retrograde (backward) and highly inclined orbit, Triton is thought to have been captured by the gravitational pull of Neptune, rather than formed in place like most other planetary moons. It is a near twin to Pluto in size and (current) distance from the Sun.

Triton's surface is very varied. The southern hemisphere is dominated by a polar ice cap, which is thought to be composed of nitrogen. In the most detailed Voyager 2 images, active geysers, ice volcanoes, were seen spewing columns of dark material many kilometres into the atmosphere. The columns were cut off at about 8km, suggesting the winds blow more strongly at that altitude.

This 1989 Voyager 2 image shows two oval cloud features rotating around Neptune. The large dark oval near the right edge revolves around the planet every 18 hours.

Pluto 76–77 ►

Gravitation 90–93 ►

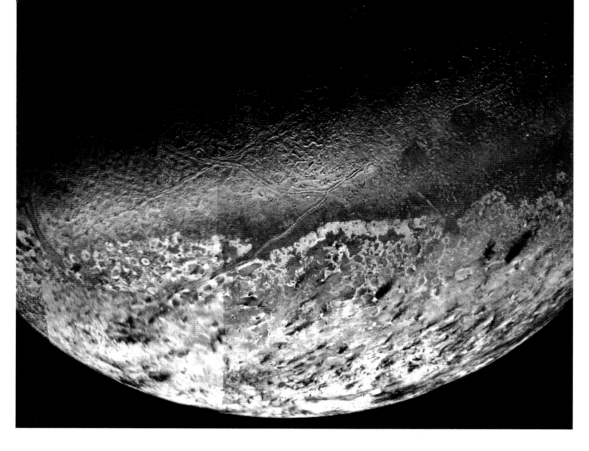

This shows the southern hemisphere of Neptune's largest moon, Triton. The large, pinkish-coloured south polar cap is at the bottom of the image. North of the cap the surface is generally darker and redder in colour. The area exhibits a plethora of unusual structural features, including long lines at the centre of the picture. The image was taken by Voyager 2 in 1989.

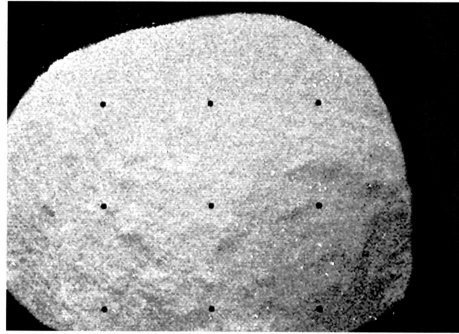

Proteus, Neptune's second largest moon, has a large feature in its southern hemisphere: a depression 250km wide and 15km deep. This image was taken by Voyager 2 in 1989

that is in vapour pressure equilibrium with the nitrogen ice covering Triton's surface, which means their respective quantities remain about level. However, due to the highly inclined orbit, Triton's extreme seasons may affect the atmosphere and bring about changes. Occultations seem to confirm this, suggesting that Triton's atmosphere may have doubled in size since the Voyager 2 encounter because of extra nitrogen gas evaporating from the ice on the surface.

SMALLER MOONS

Tiny Nereid was the only other Neptune moon known prior to the Voyager 2 mission. It also has an unusual orbit that is highly elliptical and tilted nearly 30°, suggesting a capture origin, like Triton. Nereid was not imaged well by Voyager 2, so little is known about it other than its rough irregular shape.

Voyager 2 discovered six additional moons around Neptune. These are all in circular prograde (forward) orbits near Neptune's equatorial plane, and probably evolved in place when the other planets were forming.

Surprisingly, one of these, Proteus, is larger than Nereid; it had not been discovered prior to the Voyager 2 mission because it is so close to Neptune. It is pock-marked and irregular in shape and a particularly large impact crater suggests that it came close to destruction in an earlier collision.

Triton's surface has relatively few impact craters, suggesting a young surface that has been modified by melting or flooding. The planet's northern hemisphere has a large region of puckered terrain that looks a bit like the surface of a cantaloupe melon.

Triton has a thin, hazy atmosphere. Since Voyager 2, occultations of stars have been the only method of detecting it. As Triton passes in front of the stars, gases in the atmosphere bring about changes in the star's spectrum, giving astronomers an idea of the composition of Triton's atmosphere. It is thought to consist primarily of nitrogen

MISSIONS TO NEPTUNE

Name	Date	Aims and achievements
Voyager 2	1989	Fly-by – closest approach was 5000km. Produced images of Neptune and its moons.

KUIPER BELT OBJECTS

Triton (and perhaps Nereid) is now thought to be a member of a new class of bodies called Kuiper belt objects. These are remnants of the early formation of the solar system: large chunks of rock and ice that never coalesced or collided with the primitive forming planets.

Most of the nearly 60 Kuiper belt objects discovered to date are small (less than 100km). Pluto, too, is now thought to be an end-member of this population. If Triton is a captured Kuiper belt object, then the Voyager 2 results provide insight into the characteristics of these bodies, about which we know very little.

CLUMPY RINGS

After the discovery of the Uranian rings during an occultation, astronomers used the same technique to search diligently for rings around Neptune. The results were puzzling: some occultations seemed to show rings clearly, but others did not. This conundrum was not resolved until the Voyager 2 encounter.

Images then showed three complete rings, but revealed that the rings were not uniform in their thickness and density. It was the thickest parts, dubbed rings arcs, that had been detected occasionally; the rest of the rings were too thin to be measured by occultations.

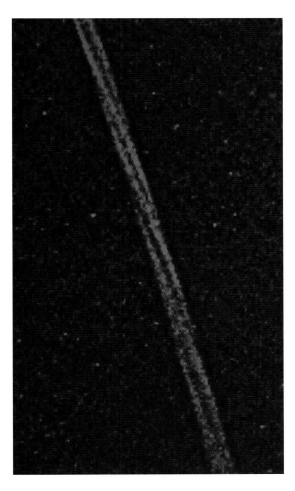

Neptune's Adams ring appears to be twisted because the original material in the rings was in clumps that formed streaks as the material orbited Neptune. This image was taken by Voyager 2 in 1989.

NEPTUNE'S SATELLITES

Name	Interesting features
Triton	Very geologically active. Orbits backward, and is probably a captured asteroid. Has a thin atmosphere, thought to consist mainly of nitrogen.
Proteus	Has a 15km deep depression that covers most of its southern hemisphere.
Nereid	Probably a captured asteroid, with a rough, irregular shape. Has a highly elliptical orbit.
Naiad, Thalassa, Despina, Galatea and Larissa	Not much is known about these satellites. They were discovered by Voyager 2 in 1989.

Astronomers are not sure what causes the rings arcs. The gravitational pulls of the moons may be partly responsible. Some of the smallest moons appear to 'shepherd' the inner edges of two rings, but no moons have been discovered in locations that would explain the shepherding mechanism. One speculation is that resonances between moons cause dust to be trapped in the arc regions.

Despite their clumpiness, Neptune's rings are circular, unlike the eccentric rings of Uranus. The three distinct rings (two rather narrow and one somewhat broader) were named Adams, Leverrier and Galle, after the astronomers who were involved in the contentious Neptune discovery saga.

There is also a broad flat sheet of dust extending from the Leverrier ring halfway out to the Adams ring.

Neptune's rings are composed of rocky material, and are significantly dustier than the rings of Uranus.

NEPTUNE TODAY

Recent Hubble Space Telescope images have continued to show remarkable changes in Neptune's atmosphere: the Great Dark Spot discovered by Voyager in 1989 had disappeared by 1994, and a new Great Dark Spot has developed in the northern hemisphere (in contrast, Jupiter's Great Red Spot has been visible for hundreds of years).

The banded structure, the parallel layers in the outer atmosphere, and cloud locations on Neptune have changed noticeably: the dark band in the southern hemisphere is distinctly narrower; some latitudes that had no clouds before are riddled with bright features, while other latitudes that had been active are now calm.

In the last 20 years a long-term study of Neptune's brightness indicates that the planet is brighter now than it ever has been. Some scientists speculate that this is related to the changing ultraviolet radiation level in sunlight as the Sun gets older, but the mechanism is not clear.

From the dynamics of Neptune's clouds, to the expanding Triton atmosphere, to the forces creating the clumpy rings, there are still many interesting puzzles to be solved in the Neptune system.

HISTORY OF DISCOVERY
The discovery of Neptune was a mathematical triumph, but a political nightmare.

After the discovery of Uranus, astronomers monitored its orbit carefully and discovered that small deviations could only be accounted for by the presence of another planet. In England, John Adams made meticulous calculations of the unknown planet's position, but for a variety of reasons he was unable to convince the Astronomer Royal to pursue the observational search, and his work languished.

Meanwhile in France, Urbain Leverrier had independently determined the new planet's orbit, although no observational programmes were begun to search for the planet.

After Leverrier's work was published, the English Astronomer Royal realized that Adam's work warranted a more serious look. By then, the French astronomer had sent his prediction to observers in Berlin. Almost immediately, the German astronomer Johann Galle discovered Neptune in its predicted location. For years, debates raged across national boundaries over who deserved credit for the discovery of Neptune. Now we acknowledge both Leverrier and Adams for the prediction, and recognize Galle for the actual observation.

PLUTO

PLUTO, THE SMALLEST AND MOST DISTANT KNOWN PLANET, HAS AN EXTREMELY ELLIPTICAL ORBIT AND AN UNUSUAL MOON, CHARON.

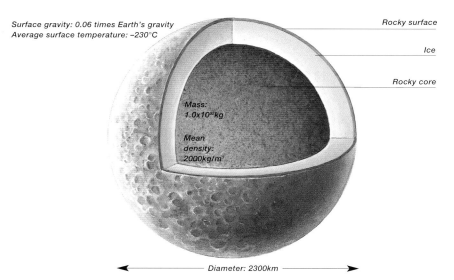

Surface gravity: 0.06 times Earth's gravity
Average surface temperature: −230°C

Rocky surface

Ice

Rocky core

Mass:
1.0x10²²kg

Mean
density:
2000kg/m³

◄———— Diameter: 2300km ————►

VITAL STATISTICS

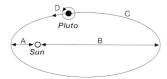

A: 4,447,000,000km
B: 7,380,000,000km
C: orbits in 248.6 Earth years
D: rotates in 6.39 Earth days

Earth

O Pluto

Pluto's diameter is 0.18 times
the size of the Earth's

ecliptic

orbit   equator

Angle of equator to
ecliptic: 112°
Angle of orbit to ecliptic: 17.1°

Satellites: one

◄ 16–19 Formation of the solar system

◄ 36 The Earth's atmosphere

◄ 40 The Moon: a small planet

Pluto's tiny size and rocky composition make it a misfit among the great gas giants in the far reaches of our solar system.

PLUTO'S ELLIPTICAL ORBIT

Pluto orbits the Sun on a long, elliptical orbit that takes 248 years to complete. Unlike other planets, Pluto's orbit is highly inclined to the plane of the Earth's orbit

Below: These two photographs of Pluto were taken on January 23 and 29 in 1930. The arrows indicate Pluto's position and show how the small planet moves against the background of the stars.

(by 17°), and carries it across a wide range of distances (from 30 to 50AU). In fact, Pluto's orbit briefly takes it nearer to the Sun than Neptune for about 20 Earth years every time the planet is near its perihelion, its closest point to the Sun. Pluto reached its most recent perihelion in 1989.

PLUTO, THE ELUSIVE PLANET

Pluto is the most difficult planet in our solar system to study because of its great distance from the Sun, and its small size. Pluto is thousands of times fainter than Jupiter, and hundreds of times fainter than Neptune, which itself requires a telescope to see. Pluto's size on our sky is less than 1/36,000 of a degree across, or the equivalent of about the size of walnut at a range of 50km.

Because of its distance from us and its faintness, astronomers could learn little about Pluto until the mid-1970s. In fact, before 1975, all that was known about the physical details of the ninth planet was that its surface colour is a bit red, that it rotates on its axis in 6.39 days and that the planet's rotation axis is tipped very far from its orbital plane, like that of Uranus.

However, as a result of the rapid advances in astronomical instrumentation, a great deal has been learned about Pluto since 1975. Astronomers now know its size and shape: it has a diameter of about 2300km and is essentially round.

COMPOSITION AND ATMOSPHERE

Also, in the late 1970s astronomers discovered that the surface of Pluto is covered in an exotic ice, methane, that only forms at extremely low temperatures. Later, in the 1990s, it was discovered that two other low-temperature ices, nitrogen ice and carbon monoxide ice, are also present in large quantities. In fact, nitrogen and carbon monoxide ices dominate Pluto's surface composition. These ices are bright, much like snow, and give Pluto its very high reflectivity.

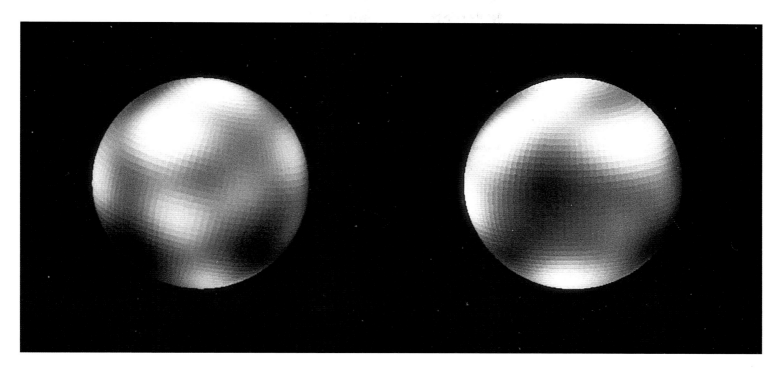

PLUTO'S SATELLITE

Name	Interesting features
Charon	At nearly half the size of Pluto, Charon is the largest moon in the solar system in relation to its parent planet. Little is known about its composition and density.

Above: These computer enhanced Hubble Space Telescope images taken in 1994 show Pluto's rotation.

Left: This is the clearest view of Pluto and Charon, taken by the Hubble Space Telescope in 1994 and enhanced in false colour.

HISTORY OF DISCOVERY
Pluto was discovered by Clyde W Tombaugh of Lowell Observatory in Flagstaff, Arizona. Tombaugh's find, on February 18, 1930, ended a series of searches for this planet that were first undertaken by Lowell's late founder, Percival Lowell, in 1905. Pluto is named after the mythological god of the underworld, and its name also has Percival Lowell's initials as its first two letters.

It is known that Pluto has an average density about twice that of water ice, or about 2000kg/m³. This implies that Pluto's interior is primarily made up of rock, but it is also thought to contain about 30 per cent water ice.

Pluto has an atmosphere that is primarily made up of nitrogen gas (like that of the Earth), with traces of carbon monoxide, methane and other gases. The atmospheric pressure at the surface of Pluto varies as the planet warms and cools during the course of its highly eccentric orbit, but never exceeds more than about 10 microbars, that is to say, about 1/100,000 of the atmospheric pressure on Earth at sea-level.

CHARON

In 1978, James Christy (with assistance from Robert Harrington) of the US Naval Observatory discovered that Pluto has a large satellite, which was named Charon after the mythical boatman who ferried souls across the river Styx to Hades.

Charon orbits over Pluto's equator at an altitude of about 17,000km in exactly the same period of time that Pluto turns on its axis. Charon therefore always hangs over one spot on Pluto.

Charon, which has a diameter of a little more than 1200km, is nearly half the size of Pluto. Astronomers do not know a great deal about its density and interior composition but they do know that its surface is darker

than that of Pluto, reflecting about only 35 per cent of the light that falls on it. Charon's surface contains a large amount of water ice, and has little colour. No proof has been found to suggest that there is any sort of atmosphere surrounding Charon.

Astronomers believe that the best explanation for the double-planet nature of the Pluto-Charon system is that Charon was created as a result of a giant collision between Pluto and another small planet in the ancient past. This scenario is much like the most widely accepted formation scenario for our own Earth-Moon system.

KUIPER BELT OBJECTS

Pluto's origin was long a mystery because of its small size compared to the other planets. This mystery was compounded by Pluto's highly unusual orbit. Pluto seemed out of place in the outer solar system. However, in recent years it has become apparent that Pluto is not alone. Instead, it orbits within a swarm of tens of thousands of still-smaller worlds orbiting beyond Neptune.

These miniature planets (and the billions of comets that orbit with them) in the so-called Kuiper belt have shown that Pluto formed not in isolation but is the largest remaining relic left over from the formation of the solar system.

Kuiper belt objects 78–79, 85 ►

Gravitation 90–93 ►

Observing the planets 160–161 ►

LEFTOVERS

THE SOLAR SYSTEM FORMED ABOUT 4.6 BILLION YEARS AGO FROM A HUGE CLOUD OF DUST AND GAS. DEBRIS LEFT OVER FROM THE FORMATION CAN STILL BE FOUND.

Just as the make-up of the planets is dictated by where they formed within the solar system, the position that the debris occupies dictates what it is made of.

ASTEROIDS

Asteroids are irregular, rocky bodies that vary in size from the largest, Ceres, at 933km to objects less than 1km across. Most asteroids are made up of primitive matter from the original solar nebula.

They lie mainly in the inner solar system among the rocky terrestrial planets; 95 per cent of them are found in the main asteroid belt between the orbits of Mars and Jupiter. Some, on the outer edge of the belt, appear to have a metallic or silicon-rich surface, which is what astronomers would expect at that distance from the Sun.

COMETS

Comets are rocky and icy bodies that can produce magnificent tails of dust and gas if they approach the Sun. The rocky, icy nucleus that forms the body of the comet is generally around a few kilometres across; the tails can stretch for as much as 10 million kilometres.

Comets are believed to exist in the outer solar system, some in the Kuiper belt which lies beyond the orbit of Neptune, and some in the Oort cloud, a large reservoir of this icy debris which surrounds the entire solar system at distances of between 30,000 and 100,000AU.

DUST AND DEBRIS

In addition to the larger debris like asteroids and comets, myriad tiny fragments pervade the entire solar system. We see some of this dust as meteors or shooting stars, brief flashes of light appearing at night as they burn up in the Earth's atmosphere. If a particle of debris is large enough to survive its fiery passage through the atmosphere it may hit the Earth's surface. It is then called a meteorite.

Other particles of dust can be seen above the Sun before dawn or after dusk, shining as the ghostly Zodiacal light or as the gegenschein.

Hale-Bopp was a very bright naked-eye comet that had its closest approach to the Sun in 1997. It is a long period comet that probably originated in the Oort cloud.

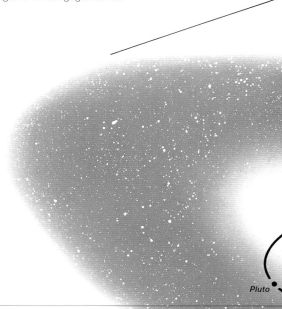

The outer solar system contains the four gas giants plus Pluto, which may be a large Kuiper belt object.

Pluto

◀ 16–19 Formation of the solar system

Gaspra is a main belt asteroid lying between the orbits of Mars and Jupiter.

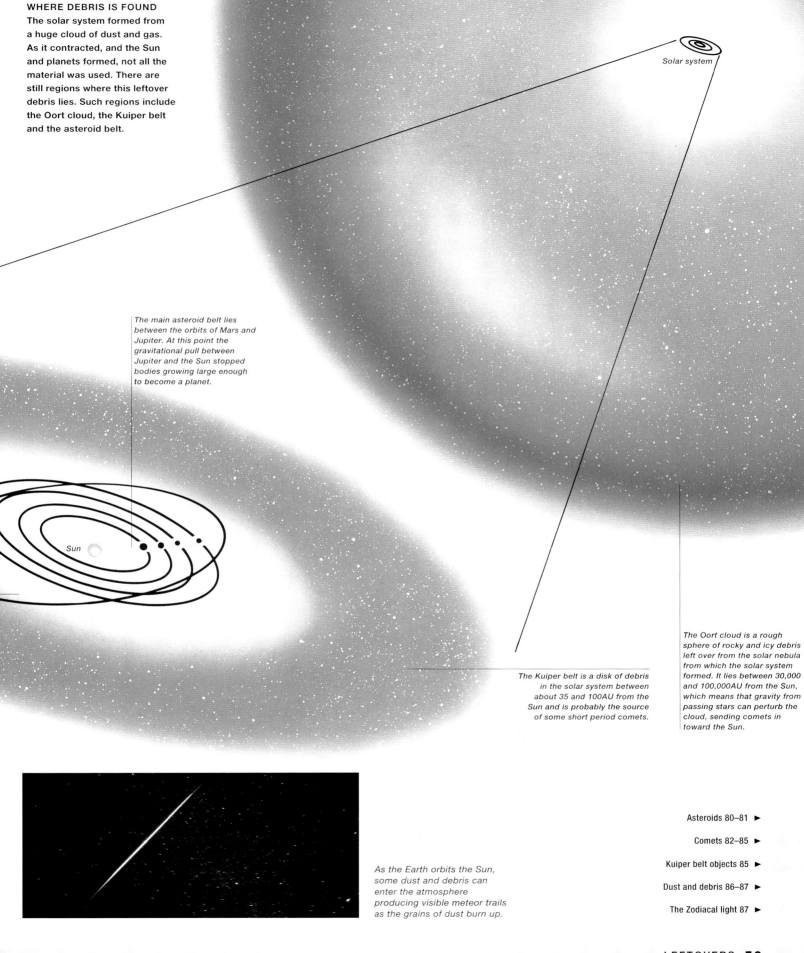

WHERE DEBRIS IS FOUND
The solar system formed from a huge cloud of dust and gas. As it contracted, and the Sun and planets formed, not all the material was used. There are still regions where this leftover debris lies. Such regions include the Oort cloud, the Kuiper belt and the asteroid belt.

Solar system

The main asteroid belt lies between the orbits of Mars and Jupiter. At this point the gravitational pull between Jupiter and the Sun stopped bodies growing large enough to become a planet.

Sun

The Kuiper belt is a disk of debris in the solar system between about 35 and 100AU from the Sun and is probably the source of some short period comets.

The Oort cloud is a rough sphere of rocky and icy debris left over from the solar nebula from which the solar system formed. It lies between 30,000 and 100,000AU from the Sun, which means that gravity from passing stars can perturb the cloud, sending comets in toward the Sun.

As the Earth orbits the Sun, some dust and debris can enter the atmosphere producing visible meteor trails as the grains of dust burn up.

ASTEROIDS

SINCE THE BEGINNING OF THE 19TH CENTURY,
ASTRONOMERS HAVE CATALOGUED MORE THAN
8000 ASTEROIDS ORBITING OUR SUN.

NAMING ASTEROIDS
Once an asteroid is found, it is tracked for months or years so its orbit can be calculated. It is then given an number and named by the discoverer. Many have been given mythological names, but some have been named after astronomers, their parents or even the observatory cat! Asteroids are the only objects in space that can be named by amateur astronomers.

The largest known asteroid, called Ceres, is two-fifths the size of Pluto, so some people refer to them as minor planets. Ceres was also the first asteroid discovered, by the Italian astronomer Giuseppe Piazzi in 1801.

ASTEROID ORBITS
The main belt asteroids are not smoothly distributed in a cloud between Mars and Jupiter. Rather, there are gaps in the main belt where very few asteroids exist. These were discovered by an American astronomer, Daniel Kirkwood, and are known as the Kirkwood gaps. They mark places where the orbital period would be a simple fraction of Jupiter's. For example, an asteroid orbiting the Sun at a distance of 375 million kilometres would complete exactly

three orbits while Jupiter orbited the Sun once. Therefore it would feel a gravitational tug from Jupiter, away from the Sun, every orbit, and quickly be moved out of that position.

However, in some places more asteroids are seen than expected: one such place is Jupiter's orbit. A swarm of a few hundred asteroids is found 60° ahead of and behind Jupiter in the same orbit. They are known as the Trojans and, as they orbit the Sun at the same rate as Jupiter, they hardly ever come close enough to the planet for their orbits to be disturbed.

Within the main belt of asteroids, some swarms can also be found. These are known as asteroid families, and are formed when two larger asteroids collide. Astronomers then see the resulting fragments as many smaller asteroids sharing similar orbits around the Sun.

ASTEROID COMPOSITIONS
No asteroid has ever been shown to have an atmosphere, so the light we see must simply be sunlight reflected from the surface. To find out what asteroids are made of, astronomers perform spectroscopy of this reflected light. Surprisingly, it turns out that an asteroid's composition depends on its distance from the Sun.

In the inner main belt nearest Mars, asteroids are made of silicate rocks (minerals containing silicon and oxygen) similar to those found on Earth. These are called 'S-type' asteroids. In the middle of the belt mostly 'C-type' asteroids are found. These appear to have rocks containing carbon, similar to some types of meteorites found on Earth.

The outer belt has asteroids that are so dark they only reflect 5 per cent of the sunlight that reaches them, and are very red. Our best guess about these 'D-type' asteroids is that there is a large amount of ices such as water ice and frozen carbon monoxide mixed in with the rock, and that charged particles from the solar wind hitting them have created chemical reactions to form the dark red colour.

This change in asteroid make-up makes sense if they were formed at the beginning of the solar system, as this change in composition fits in with theories about how the planets formed. In addition, since their formation, asteroids nearer the Sun have been heated more than those farther out, so that, over time, more ice melted and escaped. Farther away, lower temperatures mean that fewer of the ices have melted.

ASTEROID ROTATION AND SHAPES
Measuring the changes in how much light asteroids reflect shows they are not spherical like planets. This light normally rises and falls depending on how much surface of the asteroid is facing us. It is like holding a white pencil in front of a lamp. As you spin it, the amount of light reflected towards you varies depending on how much is facing you. This variation tells us that asteroids spin, but are elongated and lumpy rather than spherical. The average rotation period of an asteroid is eight hours, but Florentina takes only three hours to spin once, while Mathilde takes 17 days.

There are several ways of determining the shape of an asteroid. First, if it passes in front of a star (an occultation), measuring how long the star winks out for as it passes behind the asteroid can tell you the asteroid's size in one

Above: The Galileo spacecraft imaged the asteroid Ida on its way to Jupiter. In 1993, Galileo discovered a small moon orbiting Ida (seen here to the right of Ida).

Right: This moon, named Dactyl, is an irregular shaped body about 1.2x1.4x1km that orbits Ida at a distance of about 100km.

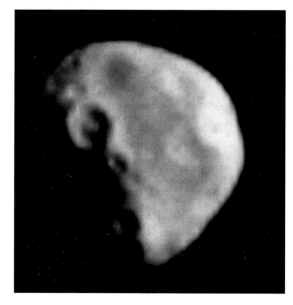

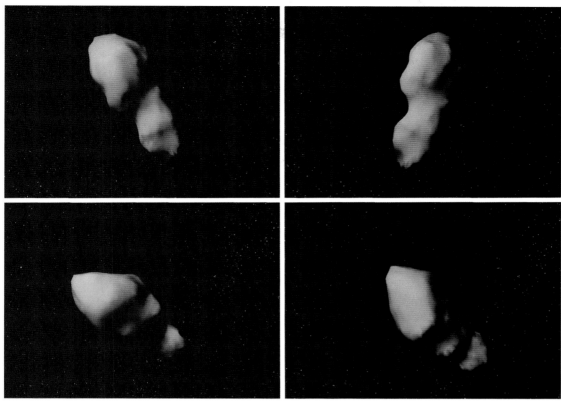

HOW TO FIND AN ASTEROID
Even the largest asteroid only appears as a star-like point of light to telescopes on Earth. In the same way as the planets, as asteroids orbit the Sun they can be seen to move against the background stars. So the easiest method of finding asteroids is by taking pictures of the sky while tracking the movement of the stars. Any asteroids will then reveal themselves by leaving short trails, while the stars appear as normal points of light.

Toutatis has shallow craters and linear ridges as revealed by this computer model created from radar data. Toutatis' orbit crosses that of the Earth in an unusual tumbling motion (see below).

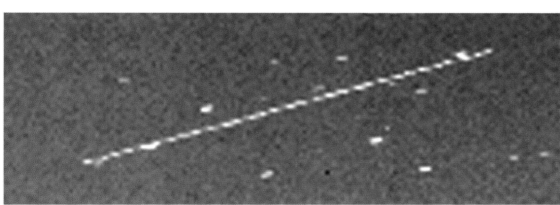

An image of asteroid Toutatis taken by the 0.9m Schmidt telescope in 1989 clearly shows Toutatis moving against the background stars. The variation in light is due to its tumbling motion as different sides of the asteroid reflect light from the Sun.

direction. By doing this several times you can build up a picture of the asteroid. Secondly, astronomers using large radio telescopes have used radar to make pictures of several near-Earth asteroids (this is the name given to asteroids with a chance of hitting the Earth). The Hubble Space Telescope has also imaged the largest asteroids in the main belt and it has, for example, found an enormous crater on asteroid Vesta.

Of course, the best way to observe an asteroid is by getting close to it with a spacecraft. On its way to Jupiter, the spacecraft Galileo passed near several asteroids. We currently have close-range pictures of asteroids Gaspra, Ida and Mathilde. Gaspra looks as expected, covered in small craters where small asteroids and meteorites have hit it. Ida surprised everyone by having a small moon (named Dactyl) orbiting it. Mathilde looked peculiar because of the huge craters on its surface.

As further pictures of asteroids are taken by spacecraft, we can expect more surprises in the future.

A mosaic of asteroid Mathilde created from images taken by the NEAR spacecraft in 1997. Several craters can be seen including the deeply shadowed one at the centre, which is about 10km deep.

Gravitation 90–93 ►

Spectroscopy 142 ►

Hubble's impact 144–145 ►

COMETS

THE SIGHT OF A BRIGHT COMET IS AWE-INSPIRING.
THROUGHOUT HISTORY, THE ARRIVAL OF THESE
BEAUTIFUL OBJECTS HAS CAUSED WONDER AND FEAR.

**DISCOVERING
AND NAMING COMETS**
It was Edmund Halley
(1656–1742) who in 1695
first calculated the orbits of
several comets around the
Sun. He found that one comet
in particular returned to the
vicinity of the Sun every 75 or
76 years. This comet has
become known as Halley's
comet. Since then, astronomers
have found over 600 comets
orbiting the Sun. Unlike the
choice about the name that is
given to asteroid hunters,
comets are always named
after their discoverers.

Yet the 20th century saw many of the comets' mysteries
uncovered, and astronomers now understand comets
well. But the first steps were taken over three centuries ago.

THE NUCLEUS

A comet has three main components. At the heart of every
comet is the nucleus, a solid mass of ice that also contains
small solid particles called 'dust'. This is not the dust that
you see in houses, but rather small bits of rock ranging from
a few hundred atoms across to boulders a few metres in
diameter. Most comet nuclei are only between 1 and 10km
across, although they can reach 100km.

Most nuclei are covered in a thin crust of icy dust.
Observations show that they are dark, reflecting only four
per cent of the sunlight that falls on their surfaces so they
are very difficult to detect when distant from the Sun. For
example, when the nucleus of Halley's comet is at the
distance of Jupiter, it is 400,000 times dimmer than the
faintest star you can see on a dark night. Therefore powerful
telescopes are needed to study the nucleus of a comet.

THE COMA

As comets move closer to the Sun, the temperature rises.
The nucleus is in a vacuum, so the ices go directly to a gas.
The effect is like putting ice in your kettle and turning it to
steam, without getting water in between.

The gas streams away from the sunlit side of the nucleus,
pushing along the smaller dust particles. This gas and dust
expand outward to form the comet's atmosphere, or coma.
The coma can reach one million kilometres across, or more.

*A Hubble Space Telescope, taken in 1996, shows the
nucleus of comet Hyakutake when the comet was 15 million
kilometres away. The image was taken through a red filter
and shows the bright area on the side facing the Sun.*

As the gases in the coma originally came from ices within
the nucleus, by studying these gases astronomers can find
out what the nucleus is made of. Spectroscopy of the
coma of a bright comet can reveal many gases; in comet
Hale-Bopp 38 types of gas were identified.

We now know that over 80 per cent of the ice in a comet
is simple water ice, similar to what is found in your freezer.
The nucleus of Halley's comet contains some 300,000
million tonnes of water ice. Another 10 per cent or more is
frozen carbon dioxide and carbon monoxide, with all the
other gases being present in smaller amounts.

COMET TAILS

The most fascinating aspect of a bright comet is its tails,
each of which forms in a different way. It is easiest to see
the tails when the comet is travelling sideways on to us.

When dust is released into the coma, it immediately
feels the effect of the sunlight hitting the dust grains and

*This optical image shows comet
Hyakutake, one of the brightest
comets to appear in the sky in recent
years. Its ion tail streams out straight
behind and often appears bluish
while its dust tail appears curved
and yellowish.*

◄ 19 Formation of a comet

◄ 21 The corona and solar wind

pushing them away from the Sun. The same thing happens to everything that is exposed to sunlight, such as your hand, but because the dust gains are so light, each of them amassing only 0.00000000001g, they are easily pushed out of the coma and away from the Sun to form the dust tail of the comet.

The dust tail of a comet can easily reach a length of 10 million kilometres. It generally appears white or yellowish, as the dust particles simply reflect sunlight, and so the dust tail appears the same colour.

THE GAS OR ION TAIL

The gas or ion tail of a comet is formed in another way. If gas molecules and atoms in the coma lose an electron, they become charged (when they are called ions) and are caught up in the solar wind, which streams away from the Sun at speeds of 400km/sec or more.

The ions are accelerated away from the coma to form the ion tail, which can extend over 100 million kilometres. In photographs this tail often appears blue, as the ionized carbon monoxide molecules in it glow strongly in blue light.

HALE-BOPP: A NEW TAIL

In 1997 astronomers studying comet Hale-Bopp found a third tail: a long straight neutral gas tail of sodium atoms stretching 10 million kilometres. The sodium atoms react to radiation pressure just as dust particles do, and because they are so light, the sodium atoms are accelerated quickly to speeds measured at 200km per second.

COMET ORBITS

Comet orbits can be split into two main groups, depending only on the orbital period, that is, how large their orbit is, and thus how long it takes the comet to travel round the Sun.

INSIDE A COMET

A comet has an irregular rocky and icy nucleus. As it approaches the Sun, ices vaporize producing the coma which hides the nucleus from view. Comets can often be seen to have two tails: the gas or dust tail and the ion or plasma tail. Astronomers discovered a new type of tail, the sodium tail, when observing comet Hale-Bopp.

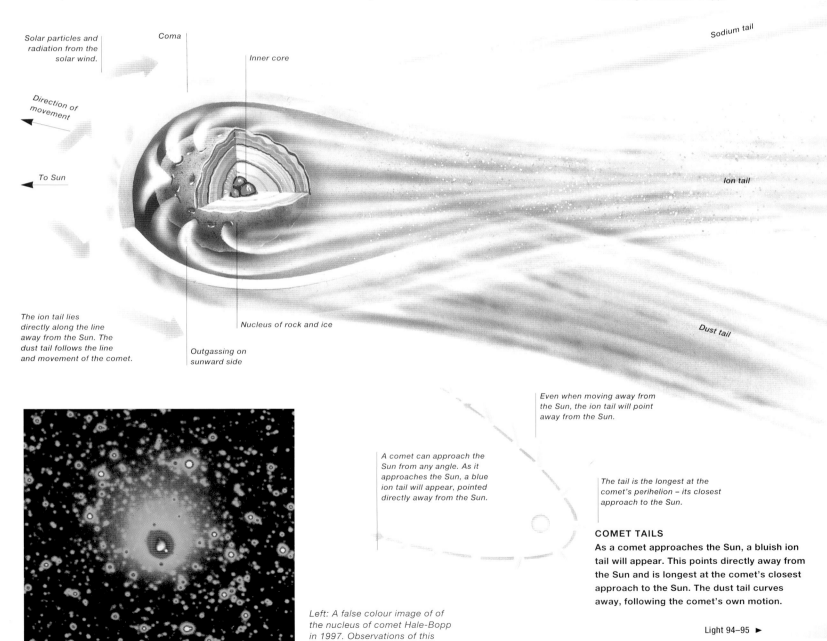

Solar particles and radiation from the solar wind.

Coma

Inner core

Direction of movement

To Sun

Sodium tail

Ion tail

The ion tail lies directly along the line away from the Sun. The dust tail follows the line and movement of the comet.

Nucleus of rock and ice

Outgassing on sunward side

Dust tail

Even when moving away from the Sun, the ion tail will point away from the Sun.

A comet can approach the Sun from any angle. As it approaches the Sun, a blue ion tail will appear, pointed directly away from the Sun.

The tail is the longest at the comet's perihelion – its closest approach to the Sun.

COMET TAILS

As a comet approaches the Sun, a bluish ion tail will appear. This points directly away from the Sun and is longest at the comet's closest approach to the Sun. The dust tail curves away, following the comet's own motion.

Light 94–95 ▶

Spectroscopy 142 ▶

Left: A false colour image of of the nucleus of comet Hale-Bopp in 1997. Observations of this comet revealed the presence of a third tail, the sodium tail.

Above: comet Swift-Tuttle. This comet has a period of 130 years and a mass of 10^{14}kg.

Right: The dark scars left by Shoemaker-Levy 9's impact with Jupiter lasted for nearly a year.

COMET SHOEMAKER-LEVY 9 Between July 16 and 24, 1994, at least 21 fragments of a comet, the smallest of them about 1km across, impacted with Jupiter and exploded in the giant planet's upper atmosphere. Astronomers think that Jupiter may have swept up many comets from the outer solar system at some stage in the past, protecting the inner planets from the catastrophic results of such collisions.

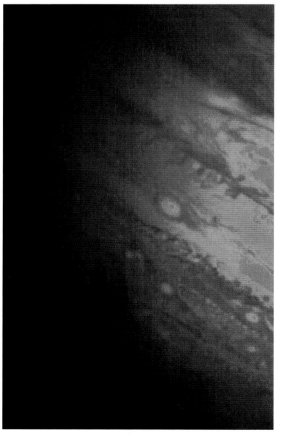

Roughly half of known comets orbit almost entirely within the orbits of Jupiter and Saturn. These take 20 years or less to go around the Sun, and are known as 'short period comets'. The comet with the smallest known orbit is comet Encke, which only takes 3.3 years to orbit the Sun.

At the other extreme, there are comets with huge orbits. Comet Hyakutake was seen in 1996, but will not be near the Sun again for another 14,000 years. Such comets are known as 'long period comets'. Astronomers have calculated that some long period comets can take up to a million years to orbit the Sun only once.

Some astronomers list a third group of intermediate comets that take anywhere between 20 and 200 years to orbit the Sun. These are known as 'Halley-type comets', after the most famous comet of all.

WHERE DO COMETS COME FROM?

Once the nature of comets became understood, a mystery opened for astronomers. The gas and dust expelled from the nucleus to form the coma and tails is lost to space, which means that a comet nucleus shrinks every time it passes the Sun – astronomers estimate that Halley's comet shrinks by about a metre every orbit. This rate of loss means that no comet can have been in its present orbit since the birth of the solar system, but they must all have spent most of their lives somewhere far away from the heat of the Sun.

In the last 50 years or so, astronomers have tracked down the sources of the comets.

THE OORT CLOUD

In 1950, Jan Oort, a Dutch astronomer, found the first comet lair when he examined the orbits of long period comets. When he plotted the maximum distance they reached from the Sun, he found that the majority reached a distance between 10,000 and 50,000AU.

As a comet slows in its orbit the farther it is from the Sun, he realized that this suggested there was a huge cloud of comets surrounding the Sun. Although astronomers cannot yet directly see comets at this distance (and will not be able to for many years), most astronomers have accepted that it exists, and it has been named the Oort cloud in honour of the discoverer.

It is believed that comets spend most of their lifetime in the Oort cloud. They will only be dislodged when a star passes near enough for its gravity to disturb comets in their orbits. In 1997 astronomers using the Hipparcos satellite found that this will next occur in roughly one million years time. When this happens comets will be sent outward into interstellar space and inward toward the Sun. The result is that this will then replenish the numbers of long period comets, replacing those that have disappeared since the last perturbation of the comet cloud.

Even though astronomers cannot see the Oort cloud directly, they can infer several properties. First, as long period comets can approach the Sun from any direction, the cloud must be roughly spherical in shape. Second, the number of comets that it contains can be estimated by calculating how many it must have lost over a period of time. Allowing for previous perturbations, the Oort cloud must contain 10 million million comets to provide the small number of long period comets that we see today.

THE KUIPER BELT

In the early 1990s telescopes became sensitive enough to try to search for comet nuclei nearer than the Oort cloud on the edge of the solar system. The first object was found there in September 1992, so dim that it was 10 million times fainter than the faintest stars seen by eye. This belt of comets is now called the Kuiper belt, in honour of one of the astronomers who predicted its existence.

Astronomers have now found over 60 Kuiper belt objects orbiting farther from the Sun than Neptune and Pluto, taking between 160 years and 720 years to orbit the Sun. Even with modern telescopes the smallest object definitely seen so far is roughly 100km across, while the largest is 500km in diameter, slightly smaller than Neptune's moon Triton. Is Triton a captured Kuiper belt object? We cannot be sure. In addition, there must be many millions of objects down to the size of a normal comet nucleus that astronomers cannot yet detect.

Many of the Kuiper belt objects found take exactly one and a half times as long to orbit the Sun as Neptune does. In these orbits they can never pass close enough to Neptune to have their orbit changed by the planet's gravity, and so may stay there forever. In fact, this is the same type of orbit as Pluto's – many astronomers now think of Pluto as nothing more than the largest Kuiper belt object.

The other objects will probably pass close to Neptune one day, and may be thrown into the inner solar system to form short period comets.

Because these objects are so faint, finding out about them is very difficult. Yet astronomers already know that they are reddish in colour, just like comet nuclei. The Infrared Space Observatory and the Hubble Space Telescope are observing them, to try to measure them and to see if they have any moons orbiting them. Best of all, a mission is being planned to visit Pluto and a Kuiper belt object. As not even the Hubble Space Telescope can see them as anything other than a point of light, a spacecraft is the only chance of mapping the surface of one of these leftovers from the beginning of the solar system.

COMET CLOUDS AROUND OTHER STARS?

In 1983 the Infrared Astronomical Satellite (IRAS) discovered that many nearby stars were enveloped in large cool clouds of material, similar to the comet clouds surrounding our Sun. Optical images have revealed a disk of material orbiting the southern star beta Pictoris, similar to the Kuiper belt around our Sun, while spectroscopy has shown what appear to be large comet-like bodies much nearer the star.

FINDING KUIPER BELT OBJECTS
The same technique is used as for finding asteroids. Because of their motion relative to the Earth, Kuiper belt objects appear to move against the background stars, so two images taken on the same night will reveal their presence. However, they move only 3.5 arc seconds in an hour, which means that very precise measurements are required to determine their orbits.

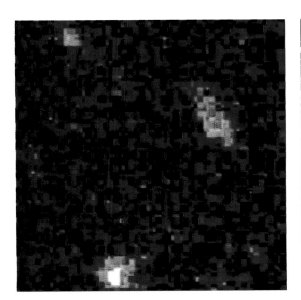
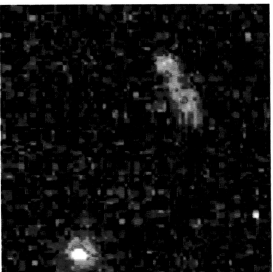

These two images show a small part of the discovery frames of 1993 SC, one of the brightest Kuiper belt objects so far discovered.

DUST AND DEBRIS

AS WELL AS PLANETS, ASTEROIDS AND COMETS, OUR SOLAR SYSTEM IS FILLED WITH EVEN SMALLER OBJECTS ORBITING THE SUN.

WHAT ARE METEORS?
The meteors seen each night are probably small fragments of rock from asteroids and comets. Meteor streams are a different matter because almost all known meteor stream orbits are shared with comets. For example, the Leonid meteor shower of November 17 shares the orbit of comet Temple-Tuttle. Meteor streams must be pieces of comet dust too massive to be swept out of the solar system.

Leonid meteors radiating out as tiny streaks against the background stars from the radiant point in the constellation of Leo.

Ranging in size from a few metres across to smaller than a grain of sand, these objects are forever being created by asteroids and comets, and being lost either to interstellar space or through collision with the Sun and planets.

METEORS

Meteors are commonly known as shooting stars. A meteor is the flash of light caused by a small piece of rock entering the Earth's atmosphere. As they travel at between 10 and 30km per second, friction with the air molecules rapidly heats the grain to thousands of degrees. You experience the same effect (but to a lesser extent!) when you rub your hands to keep warm: the friction between your hands generates heat.

The meteor literally vaporizes in a flash of light and heat. Most meteors are caused by particles the size of a grain of sand. Larger and therefore brighter meteors are known as fireballs, and can be anything from the size of a small pebble up to a large boulder.

One of the meteorites that fell near Barrell in Leicestershire, England in December 1965. It is a stony chondritic meteorite consisting of silicate materials and some nickel-iron.

Before the rock enters our atmosphere, it is following its own orbit about the Sun. The Earth regularly encounters meteor streams, where countless grains and rocks are all following similar orbits. The meteor stream always crosses the Earth's orbit at the same place because the Earth takes exactly one year to go around its orbit. As a result, we see the resulting meteor shower in the Earth's atmosphere on the same date every year.

When this occurs, it is like standing in the middle of a straight racing track, watching the cars approach you at high speed from far off, and then driving past you. Similarly, the meteors in a shower all appear to come from the same place, known as the radiant point.

The best showers occur on August 12 and December 13 each year. They are called the Perseid meteor shower and the Geminid meteor shower because their radiant points appear to be in the constellations of Perseus and Gemini respectively.

METEORITES

When a rock is large enough, it might make it through the atmosphere without being completely vaporized. In this case the remaining rock falls to the ground and is then called a meteorite. Scientists estimate that about 300,000 meteorites reach the surface of the Earth every year. Of course, many will fall in the oceans or into forests and mountain ranges. Even those that fall near towns and cities can remain undiscovered, because many meteorites look like ordinary rocks to the untrained eye.

When chemically analysed, there are many different types of meteorite. The most common finds are called chondrites, and appear to be the same type of rock that S-type asteroids are made from. Formed from normal iron and silicon bearing rocks, when cut open they contain many small circular forms called chondrules.

A much rarer type of meteorite is the carbonaceous chondrite. These can appear very like ordinary chondrites, but have large amounts of carbon in them and appear to have come from the middle of the asteroid belt.

Finally, 10 per cent of all meteorites are the stony-iron and iron-nickel meteorites, which are easy to spot as the large amounts of iron in them make them very heavy.

Meteorites are very important as they allow us to have samples of asteroids which can tell us about their history (although we can only sometimes tell which asteroid a meteorite has come from). For example, the chondrules in ordinary chondrites appear to have formed at very high

SOME ANNUAL METEOR SHOWERS

Date	Name	Number/hour
January 3	Quadrantids	60
April 22	Lyrids	10
August 12	Perseids	75
November 17	Leonids	10
December 13	Geminids	75

temperatures, and show that when the meteorites first formed at the same time as the rest of the planets, parts of the nebula around the Sun must have been very hot.

CRATERING

If the meteorite is big enough, it will create a crater in the ground. One of the most famous is the Barringer 'meteor' crater in Arizona. About 1.6km across, it was formed 49,000 years ago when a stony-iron meteorite estimated to be between 50 and 70m across (really a small asteroid) struck the Earth's surface.

More than 130 craters have so far been found on the Earth. Many more have been created and subsequently destroyed by both erosion and geological activity. You can see what our planet would have looked like without erosion, simply by looking at the Moon through binoculars. Here the craters have been preserved for us to see.

What would happen if such an impact happened today? The energy released during the impact of a small asteroid (say 1.5km across) is equivalent to all of the nuclear weapons ever created going off in a single spot. Besides the tremendous heat, enough rock and dust would be deposited in the atmosphere to change the climate all over the Earth, while if it landed near water, huge tidal waves (tsunamis) would devastate cities on the edge of the ocean all over the Earth. Indeed, it is now believed that the impact of an asteroid or comet 65 million years ago assisted in the extinction of the dinosaurs. Geologists have found the resulting 180km wide crater in the Yucatan Peninsula in Mexico. Luckily for us, such devastating impacts only happen once every 100 million years or so.

THE ZODIACAL LIGHT

Uncountable numbers of smaller particles are left behind in a cometary trail. These particles lie predominantly along the Sun's passage as seen from Earth because comets lose more material the closer they approach the Sun. The dust particles reflect the Sun's light, and sometimes this ghostly reflected light can be seen in the west after sunset, or in the east before sunrise. This light is called the Zodiacal light, with an associated phenomena, the gegenschein.

Above: The Arizona 'meteor' crater (Barringer crater) is 12km wide and 180m deep. It was formed about 25,000 years ago by the impact of a meteorite about 50–70m in diameter.

Left: The Zodiacal light is the faint glow of dust particles in the interplanetary medium scattering sunlight.

OTHER SOURCES

A few meteorites appear to have come from the Moon and Mars. It is thought that these rocks were blasted off the surface of these bodies when a large asteroid hit them, and thrown into orbits that reached the Earth. This is important for scientists studying Mars, as no spacecraft has yet sent back rocks from that planet. In 1996 scientists announced that one Martian meteorite may contain the fossilized remains of life. Although that has not been proved, it demonstrates how important the study of meteorites can be.

Orbits 90 ▶

Observing meteors 162 ▶

COLLISION COURSE

Asteroids in the main belt pose no threat to Earth, but some close-approach asteroids – known as rogues – demand special attention. Rogue asteroids get close to Earth, so close that the chance of a collision is something we can't ignore. Some even call rogues COTEs: Celestial Objects Threatening Earth.

BY DR DAVID HUGHES

Using spacecraft, astronomers have been able to obtain close-up images of some asteroids. These rocky bodies show cratered surfaces – the results of many impacts. This particular asteroid, Gaspra, lies in the main belt and will never approach Earth.

❝ The difference between rogue asteroids and asteroids that lie within the main asteroid belt is really just a matter of distance and size. Main belt asteroids are fairly harmless bodies, since they remain primarily between the orbits of Mars and Jupiter. Their nearly circular orbits are inclined at low angles to the Earth's orbital plane and they usually stay where they are. Only rarely are they perturbed into Earth-crossing orbits.

Two groups of COTE asteroids are called Atens and Apollos. Aten asteroids follow orbital paths that lie mainly inside Earth's orbit, while Apollo asteroids cross the Earth's orbit and have an average distance of 1AU from the Sun. The largest member of these groups is asteroid 1685 Toro, which is about 12km across.

Because they are relatively bright and large, we have probably now spotted all 13 or so Apollo asteroids with diameters of more than 5km. While this may sound comforting, bear in mind that a typical Earth-crossing asteroid is only 1km across; at its brightest, it has an apparent magnitude of only around +17 – nearly 16 times as faint as Pluto – unless it comes really close. The really frightening thought is that there are many more smaller, fainter asteroids: massive numbers of these must be crossing the Earth's orbit without being discovered. I estimate that when it comes to Apollos with diameters that are greater than 1km, we have actually seen just 5.5 per cent of them. Thus, even though we have recorded about 100 Earth-crossing asteroids with diameters of 1km or more, there may actually be as many as 1800 of them out there, whirling around the inner solar system in orbits that are a little too close for comfort.

If the orbit of an asteroid crosses the orbit of Earth, the two bodies have a chance of hitting each other. While you might think that hitting a smaller rogue wouldn't be such a frightening prospect, just consider what effect a collision with a 1km asteroid would have on our world. The speed of impact is obtained by combining the speeds of both bodies' orbits and adding the effect of the Earth's gravitational attraction. So if a 1km

The threat is out there, and it is all too real for comfort. Can we do anything to stop rogue asteroids hitting Earth?

asteroid (a mass of 1500 million cars) collided with Earth at nearly 21km per second, it would hit us with an explosion equivalent to 8.1 kilotons of TNT or to four million atom bombs the size of the one which devastated Hiroshima in the 1940s.

It is difficult to relate the kinetic energy – the energy associated with motion – of an asteroid to the size of the crater it could produce on the Earth's surface. We do have clues, however, and the best come from a series of small surface nuclear tests that were carried out during the Cold War. These suggest that our 1km asteroid would produce a crater with a diameter of about 26km.

We don't often witness the creation of craters of this size. It is estimated that asteroid impacts of this type are so rare that more than 100,000 human generations would pass between one and the next. To calculate the crater-production rate geologists survey old, stable regions of the Earth's surface and count the craters they contain. However, this method is fraught with problems: glaciation, sedimentation, wind and rain all erode craters, and the smaller ones disappear relatively quickly.

So it is more accurate if geologists use only large and relatively recent craters to find out how crater size distribution varies both with time and with asteroid diameter. The effects of erosion are minimized if only craters larger than 19km and less than 100,000 million years old are surveyed. Unfortunately, this method leaves us only six big craters to choose from: Montagnais (diameter 45km), Manson (35km), Mistastin Lake (28km), Steen River (25km), Haughton (20.5km), and Eagle Butte (19km). All of these have been produced in an area of about 10 million square kilometres in North America.

Based on these limitations the data indicate that craters with diameters of more than 19km are being created on the Earth at an average rate of one every 380,000 years. Craters larger than 26km (produced by 1km asteroids) are created on average every 640,000 years, while those craters larger than 1km occur every 940 years.

The worrying thing about a rogue asteroid is that its impact with Earth might produce such a

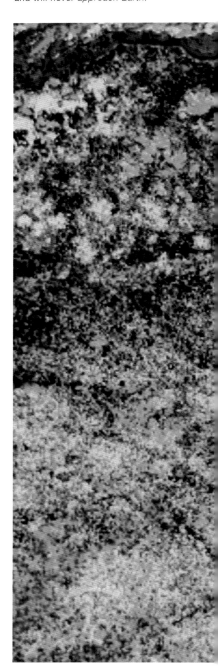

large crater that the ejected dust, together with the accompanying fire storm, seismic damage, tidal wave and acid rain could result in such major damage and climatic change that we would experience the end of civilization as we know it. Such an impact would be the equivalent of the instantaneous destruction of a country the size of the US or Brazil. The result would be the eradication of world agriculture for at least one year. Geologists estimate that a 4km rogue asteroid could do the job, since it has enough energy just before it impacts to produce a 100km crater. Fortunately, such an impact event occurs only about once every 10 million years.

Still, the threat is out there, and it is all too real for comfort. Can we do anything to stop rogue asteroids hitting Earth?

We think we have found nearly all the asteroids bigger than 4km that are approaching Earth at the present time. But frankly, this is not good enough. We need to extend this work to discover *all* the 4km asteroids that are in the main asteroid belt, simply because one of them might be pushed into an Earth-crossing orbit at some time during the next 10 million years by the influence of the major planets. It's a daunting task – especially when you consider that there are probably nine million of them. When one is discovered, astronomers should then calculate its orbit over the next 10 million years with almost unimaginable accuracy to see if it could be moved into an Earth-intersecting path. The next step should naturally be to prevent the possibility by destroying the rogue asteroid – that is, once we have the capability to do so. **"**

The Chicxulub crater, a buried impact crater more than 200km in diameter, was discovered in the Mexican state of Yucatan by NASA scientists studying satellite images for water sources used by ancient Mayan cities. The apparent age, location and size of the Yucatan impact makes it the best candidate for the global catastrophic event that caused the extinction of the dinosaurs.

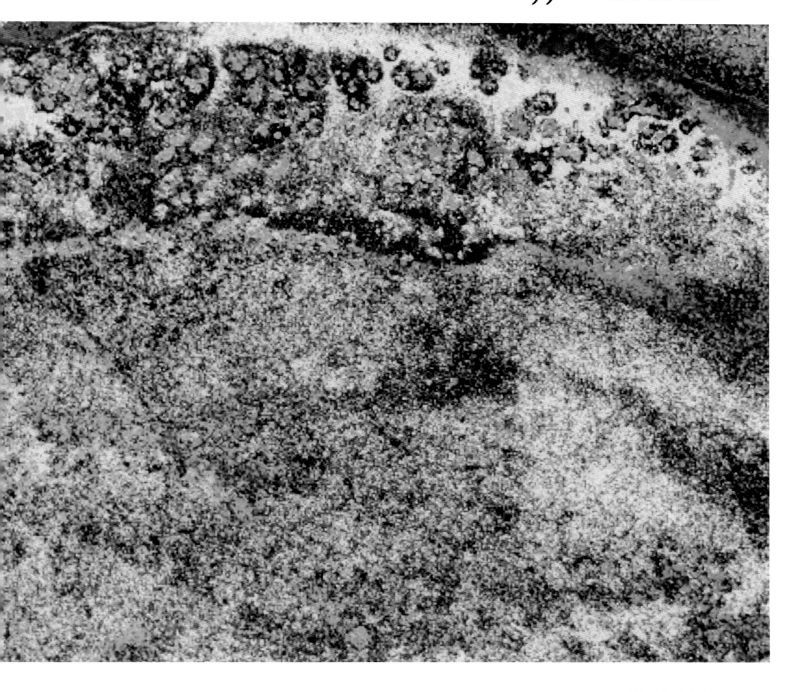

GRAVITATION

GRAVITY AFFECTS ALL MATTER IN THE UNIVERSE. IT
CAUSES AN APPLE TO FALL TO THE GROUND, JUST AS
IT HOLDS THE PLANETS IN ORBIT AROUND THE SUN.

KEPLER'S LAWS
1. Each planet orbits around
the Sun in an ellipse with the
Sun at one focus of the ellipse.
2. A line joining the Sun to a
planet sweeps out equal areas
in equal times, so the closer
a planet is to the Sun, the
faster it moves.
3. The square of a planet's
orbital period is proportional
to the cube of its mean
distance from the Sun.

Essentially, gravity is the force of attraction between all types of matter. Gravity acts between you and the Earth, holding you to the floor or the ground. Gravity is also acting between the Earth and this book, and between you and this book.

Gravity exists between all matter in the universe. It cannot be turned off, modified or blocked (although it can be counteracted by the introduction of an opposing force). In physics terms, it is one of the four fundamental forces of nature; the other three are the weak and the strong nuclear forces, and electromagnetism.

KEPLER'S AND NEWTON'S LAWS

It is the force of gravity that keeps all the planets in the solar system orbiting the Sun. The Sun contains 99.86 per cent of all the mass in the solar system; thus it is by far the most dominant player in the gravitation game.

For many years, astronomers struggled to find a model that could explain the apparent motion of the planets in the sky, in the hope that they could eventually predict planetary positions. In pursuit of this goal, the Danish astronomer Tycho Brahe made many painstaking and accurate observations of the positions of the planets from 1572 to 1596. His assistant, German mathematician Johannes Kepler, used Tycho's observations to work out

that the planets travelled in elliptical orbits round the Sun rather than circular ones as had been assumed until then.

Kepler formulated three important laws (see boxes below) concerning the motion of the planets, and these laws bear his name. Nearly 80 years later, Isaac Newton explained *why* Kepler's Laws work since they can be derived from his Law of Gravitation and his three Laws of Motion.

Newton's Universal Law of Gravitation, put forward in 1687 in his *Philosophiae Naturalis Principia Mathematica* (known as the *Principia*), shows that the force of gravitation depends only upon the masses of the bodies concerned and their distance apart. It also tells us that gravity is always attractive. In addition, gravitational force is symmetrical; this means that the force exerted on object A by object B is the same as the force exerted on object B by object A.

DISCOVERING NEPTUNE

Kepler's Laws and Newton's Law of Gravitation played an important part in the discovery of the planet Neptune. Uranus had been discovered in 1781, and observations by a number of astronomers throughout the following 60 years showed that there were irregularities in its orbit. Careful calculations of gravitational effects showed that these orbital irregularities could be explained by the gravitational pull of another, unknown planet, farther away from the Sun than Uranus.

French mathematician Urbain Leverrier calculated where this planet should be to have the effect it did on Uranus, and using Leverrier's data as a basis for his observations, German astronomer Johann Gottfried Galle did indeed identify the presence of a new planet in the predicted location in September 1846. It was named Neptune.

NEWTON'S THIRD LAW OF GRAVITATION

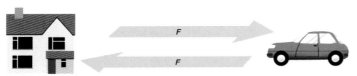

Gravity is symmetrical. The house and car exert equal gravitational force (F) on each other.

NEWTON'S UNIVERSAL LAW OF GRAVITATION

Doubling the distance between
two objects quarters the force
of gravitation.

THE UNIVERSAL LAW OF GRAVITATION
Newton's Law states that two bodies attract each other with a force that is directly proportional to the product of their masses and inversely proportional to the square of the distance between them. This means that if either mass is increased, the force increases. However, if the distance between the objects increases, the force becomes less. The Universal Law of Gravitation is an example of the 'inverse square law'.

ORBITS

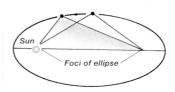

First Law: Kepler calculated that planets orbit around the Sun in ellipses, rather than circles.

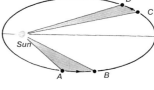

Second Law: The planet moves faster when it is closer to the Sun. It takes the same time to travel from A–B as from C–D.

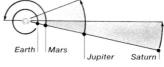

Third Law: During an Earth orbit, Mars makes just under ½ an orbit, Jupiter about 1/12 of an orbit and Saturn about 1/30 of an orbit.

The planets would travel in straight lines if the Sun did not exert force on them (Newton's First Law of Motion), but the gravitation from the Sun pulls them towards it. However, the planets would fall into the Sun if it were not for the motion they retain from the original motion of the solar nebula. The motion and the force act to keep the planets in orbit.

WEIGHT AND MASS

The concepts of weight and mass are easily confused. Mass simply refers to the amount of matter in a body. Therefore, astronauts travelling to different planets in our solar system would have the same mass regardless of which planet they were on.

Weight, however, is a measure of the gravitational force that is exerted on a body. Astronauts would have different weights on different planets due to the latter's different gravitational forces.

Astronauts who are on board spacecraft in orbit around the Earth can be seen floating: they experience weightlessness. They are, in fact, constantly falling down toward the Earth, but because of their sideways motion, they travel in an orbit around the Earth instead of falling to the ground. Obviously, they still retain their mass, but the orbit of the spacecraft, and so the astronauts, is such that they are in freefall within the Earth's gravitational field. In freefall, a body experiences no weight.

ESCAPE VELOCITY

Another important application of the Law of Gravitation concerns the idea of 'escape velocity' – that is, the minimum speed at which an object must move in order to escape another object's gravitational field.

Being able to calculate escape velocities has several important applications. For one thing, it enables scientists to work out the speed needed by a space vehicle to escape from a planet or satellite. Additionally, it helps explain why

Six astronauts on the Moon would weigh the same as one astronaut on the Earth.

some planets are able to retain an atmosphere. In order to have an atmosphere, the average speed of any atmospheric gas molecules must be well below the escape velocity of the planet concerned. The more massive the planet, the higher the escape velocity. Thus, the more massive planets are able to retain their atmospheres more easily.

THE GRAVITATIONAL SLINGSHOT

Another practical application of gravitational law that has been used in recent times is the slingshot, or gravity-assist, effect. This effect has been used to change the speed and trajectory of various interplanetary probes.

Astronauts, like these outside the Endeavour Shuttle, can experience weightlessness when in orbit about the Earth. The same is true of objects in orbit, and this means that astronauts can move massive bodies, such as the COSTAR corrective optics for the Hubble Space Telescope with ease.

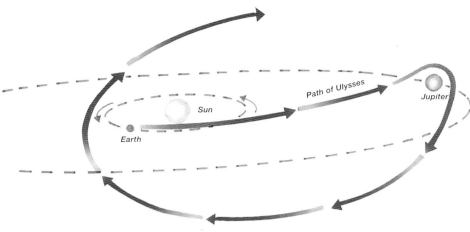

Path of Ulysses
Jupiter
Sun
Earth

GRAVITATIONAL SLINGSHOT
The gravitational pull of Jupiter deflected the Ulysses spacecraft out of the plane of the solar system so it passed by both poles of the Sun. Not only did Ulysses' passage round Jupiter change its direction, but it increased its speed.

By directing a spacecraft to pass relatively close to a planet, the spacecraft is pulled into the gravitational field of the planet, and thus changes direction. Because of its initial speed, it is not pulled into orbit around the planet, but gains a tiny fraction of the orbital energy of the planet as it flies past. The planet loses a corresponding amount of energy but this is almost unnoticeable because of its huge mass.

Examples of the use of this technique have included Giotto, the probe sent to study Halley's comet at its approach to the Sun in 1986. Giotto flew past the Earth in 1990 so that its orbit could be modified in order to encounter a second comet, Grigg-Skjellerup, in 1992. Similarly, the Ulysses spacecraft, which flew past Jupiter in 1992, had its orbit deflected to enable it to travel out of the ecliptic plane to fly over the Sun's south and north poles.

The Cassini mission, launched in 1997, will use this technique to reach Saturn in 2004. There is no rocket presently available that is powerful enough to send the massive Cassini spacecraft directly from Earth to Saturn; therefore this mission will employ no fewer than four gravity assists (two at Venus, one at Earth and one at Jupiter) in order to reach the ringed planet.

BINARY AND MULTIPLE STAR SYSTEMS

Astronomers now believe that more than half of all stars exist in binary, or even multiple, systems. In a binary system, the mutual gravitational attraction causes the two stars to revolve around each other in an astronomical waltz. They move in elliptical orbits, as predicted by Kepler's Laws, around the centre of mass – a point somewhere between the centres of the two stars.

Sometimes it's possible to observe both stars in a binary pair and to plot their individual paths (this is known as a visual binary). In other cases, only one star is observable as the other is too faint. In this case, however, the presence of the unseen star can be deduced from the 'perturbation' on the motion of the observed star – that is, the wobble in its motion due to its companion's gravitational pull.

By studying visual binaries, it is possible – using Kepler's and Newton's Laws – to determine the masses of the stars. This is important as it is the only direct method of doing so.

Stars are not the only objects that can affect the movement of another star. Planets can do the same; being less massive, however, their effect is much less – although it is potentially observable. In fact, our Sun is not quite static, but moves about because of the gravitational influence of the nine planets. Since the planets are in continual motion, the pull of them on the Sun is always changing.

If a nearby star has a system of planets like our own, the planets themselves will be far too faint to observe. Even so, their pull on their star will make it wobble, and this might well be detectable. Such phenomena have indeed been observed in recent years.

THE CURVATURE OF SPACE AND TIME

Newton's Law of Gravitation and related Laws of Motion (Newtonian mechanics), as well as laws derived from observations such as those of Kepler, have been remarkably successful in describing and explaining astronomical motions ranging from those in our own solar system to those in the most distant galaxies. However, as astronomy developed to the point where extremely accurate measurements could be made of very exotic environments – black holes, neutron stars and white dwarfs – it became clear that there were some small discrepancies that could not be explained by Newtonian mechanics.

This is where Einstein's General Theory of Relativity comes into play. Developed in 1915, this theory describes how, in the presence of a gravitational field, space becomes curved and this affects the movement of anything nearby. This only becomes noticeable in the presence of massive objects with very strong gravitational fields: otherwise the extra effects of General Relativity over and above Newtonian mechanics are not detectable.

MERCURY'S ANOMALY

The effect of the curvature of space on the motion of astronomical bodies, and also on light, in the presence of strong gravitational fields was predicted quite definitively by the General Theory of Relativity. For example, in the region around the Sun where the Sun's gravitational field is at its

BINARY STAR SYSTEM
The movement of a star against the background of more distant stars is called proper motion. If the star follows a 'wavy' motion, it may be because it has an unseen companion orbiting it and pulling it around with its gravitational interaction.

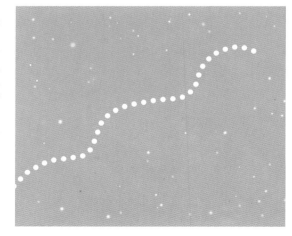

strongest, it was predicted that any planet orbiting in that region would very slowly change its orbit.

As it is the closest planet to the Sun, Mercury's orbit would show the largest effect of this curvature. If Mercury's position were plotted over a period of 100 years, its actual position would be 43 arc seconds ahead of the position predicted by Newton's Laws alone. Einstein's equations successfully accounted for this anomaly in Mercury's orbit, explaining a phenomenon that had frustrated astronomers for more than half a century.

GRAVITATIONAL LENSES

Not only does the General Theory of Relativity predict perturbations of the motion of solid bodies, it also predicts disturbances of electromagnetic radiation. An example of this is when light from a star passes close to the edge of the Sun. The Sun's gravitational field 'bends' or warps the light toward itself, making the star appear in a different position than it would if the Sun were not there. This has now been observed, and the deflection of light is almost exactly as predicted by the theory.

A more exotic illustration of the bending of light occurs when certain distant galaxies or quasars are observed. If the light from the distant object passes close to an intervening galaxy or cluster of galaxies, the light is bent in such a way that the intervening object acts as an optical lens, magnifying the image. The effect of this can be to produce two or more identical images of the original object.The effect is a bit like seeing a car in a curved mirror round a sharp bend in the road before you see the actual car. The large cluster of galaxies acts like the mirror by bending the light from the more distant galaxy.

The thin blue smudges around the cluster of galaxies in the centre of this HST picture are images of another galaxy beyond. The gravity of the large cluster warps and breaks up the light of the distant galaxy so that the image is displaced and distorted. It is an example of a gravitational lens.

LIGHT

LIGHT IS A TYPE OF ENERGY. IT IS THE PART THAT
OUR EYES CAN DETECT OF A WIDER MANIFESTATION
KNOWN AS ELECTROMAGNETIC RADIATION.

ISAAC NEWTON
In the late 1600s Isaac Newton (1642–1727) investigated the 'phenomenon of colours', the spectrum produced after passing sunlight through a prism. Newton concluded that white light was made up of all the colours, instead of the colours being added by the prism as previously thought. He concluded this by passing the spectrum of light though another prism, thereby showing it turned back into the original beam of white light.

WAVELENGTH
All electromagnetic radiation moves through space as a wave. The energy depends on the wavelength: the shorter the wavelength, the higher the energy of the radiation.

In addition to visible light, other types of electromagnetic radiation include X-rays, radio waves, ultraviolet and infrared radiation. All of these forms of energy make up what is called the electromagnetic spectrum, yet the only fundamental differences between one type of radiation and another are frequency and wavelength.

FREQUENCY AND WAVELENGTH

Imagine you are on a beach, and you're counting how many waves hit the shore during a certain period of time. This measurement is known as frequency. For example, if 10 waves hit the shore in the period of one minute, the frequency is 10 waves per minute.

Now imagine measuring the distance between the crest of each wave. This measurement is known as wavelength. Thus, if this distance is 1m, the distance across the 10 waves above would be 10m, which means the sea would be moving at the rate of 10m per minute. In other words, the wave velocity, or speed, can be found by multiplying the frequency by the wavelength.

This shows that two waves could have the same velocity even if one had a high frequency and a short wavelength while the other had a low frequency and a long wavelength. If waves were hitting the shore at the rate of only one per minute, but the distance between their wave crests was 10m, then the sea would still be moving at the rate of 10m/min – the same velocity as in the first example.

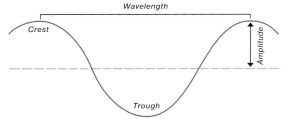

A SPECTRUM OF COLOURS

A rainbow is a familiar example of a visible spectrum. It occurs when sunlight is broken up into its constituent colours by water droplets present in the atmosphere. With a spectroscope any source of light can be made to reveal its spectrum, and the spectra of many objects pass well beyond visible white light and range right across the whole of the electromagnetic spectrum.

THE ELECTROMAGNETIC SPECTRUM

Despite the fact that we cannot see them, all forms of radiation travel through a vacuum at the same speed – the speed of light. Thus, in the same way as waves on the beach, types of radiation with short wavelengths (gamma rays, X-rays, ultraviolet radiation) must have high frequencies. Conversely, types of radiation with long wavelengths (infrared radiation and radio waves) must have low frequencies.

THE DOPPLER EFFECT

Whenever the source of electromagnetic radiation moves in relation to the 'receiver' of such radiation, a phenomenon occurs which is known as the Doppler effect. The relative motion between the two objects causes an apparent change in the frequency of radiation.

The Doppler effect also happens with sound waves. Think of a train blowing its whistle as it passes you. As the train approaches, the sound waves generated by its whistle are bunched up and reach you with a greater frequency (and thus at a higher pitch) than when the train is moving away from you (when the sound waves are stretched out and reach you with a lower frequency). The result is a dramatic drop in pitch of the whistle as the train passes you by.

In the case of light, one of the best-known examples of the Doppler effect can be seen when a distant galaxy is found to be receding from us. The light seen has a longer wavelength than would be the case if the galaxy were stationary. The longer the wavelength, the redder the light appears, so this phenomenon is referred to as the redshift.

PACKETS OF ENERGY

The wave picture of the nature of light allows a variety of related phenomena associated with electromagnetic radiation to be explained. However, some phenomena can be explained only by considering radiation as being

REDSHIFT AND BLUESHIFT
An object is observed by the radiation it emits or reflects into a detector. If the object is stationary with respect to the observer, the emission and absorption lines in its spectrum will be in certain places – at certain wavelengths.

If the object is approaching the observer, the radiation is 'bunched up'. Its frequency increases and its wavelength decreases. The absorption or emission lines will move towards the blue end of the spectrum. They are 'blueshifted'.

Similarly, if the object is moving away from the observer, the wavelength of the radiation is 'stretched' to longer wavelengths and lines within the spectrum are 'redshifted'.

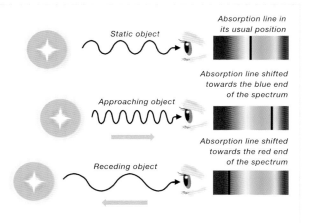

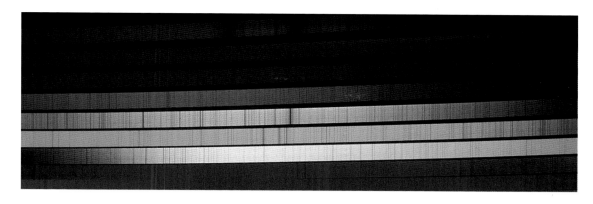

The solar spectrum has dark absorption lines, first studied in detail in 1814 by Joseph von Fraunhofer. More than 25,000 lines have now been identified within the spectrum, giving a great deal of information about the chemical composition of the Sun.

made up of a beam of particles. This concept is called the wave-particle duality of light.

Radiation can be thought of as packets of energy which travel through space as waves, but behave as particles when interacting with matter. The packets, or quanta, of energy are called photons. The energy of the photons is directly related to the frequency of the radiation and the pressure exerted by the photons is called radiative pressure.

ATOMIC SPECTRA

An atom consists of a nucleus (protons and neutrons) with a number of electrons in orbit around it. Only certain electron orbits are possible, with each orbit corresponding to a different 'level', or amount, of energy. If an electron drops from a higher to a lower energy orbit, any energy that is lost is transformed into a particle of radiation, a photon; thus, all atoms can emit radiation. Similarly, atoms can absorb photons of energy, which causes electrons to rise to a higher energy level.

Because of the different electron orbits, each particular chemical element (hydrogen, nitrogen, etc) absorbs and emits photons of different amounts of energy. Since the photons have different energy, they will also have different wavelengths; an atom of hydrogen, for instance, will always give off the same energy 'signature', and it will always be different from that given off by an atom of helium. For this reason, looking at the spectrum of a star allows us to discover its chemical composition.

We can also discover the chemical make-up of any material that may lie between us and an object. Radiation wavelengths emitted by the source will be absorbed by the intervening material. The elements within the material absorb only at specific wavelengths because of the electrons in their orbits. This produces dark absorption bands within the spectrum.

Similarly, if there is a hot source nearby, such as an incandescent cloud of gas that is emitting photons, it will produce a different type of spectrum, made up of individual bright lines. This is known as an emission spectrum and each of the lines (called emission lines) corresponds to a particular element or group of elements.

Spectroscopy, the analysis of objects by studying the radiation they absorb or emit, has developed significantly over the past century, so that now even minute traces of a material can be detected and identified. It is now the main technique by which astronomers study the conditions in the most distant objects in our own galaxy and beyond.

ATOMIC SPECTRA, ABSORPTION AND EMISSION LINES

Atoms can absorb or emit photons of radiation, creating absorption or emission lines in spectra. The wavelengths of the lines can indicate the composition of the object.

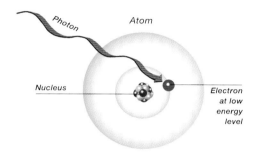

Electrons can only exist in certain energy levels (or orbits). When a photon of radiation is absorbed by an electron it moves to a higher energy level.

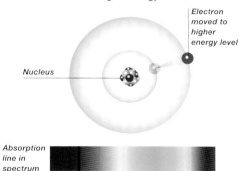

The amount of radiation absorbed by the electron creates a dark line in the spectrum – an absorption line. The position of the absorption line indicates the type of atom.

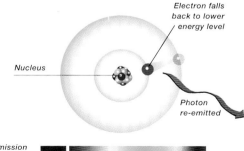

When the electron falls back to its original energy level, the photon of radiation is emitted, creating a bright emission line in the spectrum.

The electromagnetic spectrum 140–141 ►

Spectroscopy 142 ►

OTHER SOLAR SYSTEMS?

The existence of extrasolar planets – that is, planets beyond the Sun's – is vital to astronomers for two reasons. First, planets provide a platform on which life can arise and evolve: if ours were the only solar system in the universe, ours would also be the only civilization in the universe. Second, observing other solar systems will allow astronomers to see what rules might govern all solar systems.

BY KEN CROSWELL

Above: An artist's impression of the planet around the Sun-like star, 51 Pegasi. The data suggests that the planet is the size of Saturn or Jupiter, but closer to its parent star than Mercury is in our own solar system. It is believed that it formed farther out before spiralling inward.

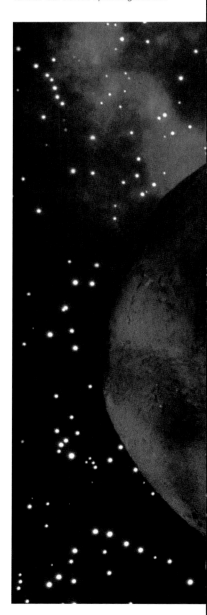

" But extrasolar planets are as difficult to detect as they are important to find. Unlike stars, planets emit no light of their own; they merely reflect the light of the star they orbit. Even the brightest planets in our sky – Venus, Mars and Jupiter – would be far too faint for the naked eye to see if they were as far away as the nearest star. Moreover, they would be lost in the star's glare. Consequently, although astronomers have long known of countless stars and galaxies, until 1991, the number of known planets stood at just nine: the nine worlds that whirl around the Sun.

Fortunately, planets do possess something that gives their presence away: mass. Anything with mass exerts gravity, so an orbiting planet's gravity causes its parent star to wobble. Since astronomers on Earth can see the star, they have a chance of detecting the star's wobble and thus inferring the planet's existence.

However, these stellar wobbles are tiny, because stars are big and planets small. In our solar system, the largest planet, Jupiter, has 318 times the mass of Earth, but that's still only $\frac{1}{1000}$ of the mass of the Sun. For many years, astronomers claimed to have detected planets by the wobbles they induced in nearby stars. Later work, however, showed that these apparent wobbles lay not in the stars, but in the telescopes that observed them!

Interpreting the data showing that there are planets orbiting round other stars can still be controversial. Alternative ideas have been put forward to explain the apparent wobbles, for example that a pulsating star could produce the same data as a planet-induced wobble.

In 1991 radio astronomers Alex Wolszczan and Dale Frail detected the first genuine extrasolar planets. The location of these planets proved to be a shock, for they revolved around a pulsar: a dense, fast-spinning star that forms when a massive star explodes as a supernova. Such an explosion should free any orbiting planets, because it reduces the star's mass and gravitational

If astronomers detect large amounts of oxygen in another planet's atmosphere, that planet might well have life.

pull. Nevertheless, at least three planets orbit around the pulsar PSR B1257+12, in Virgo.

Like a lighthouse, a pulsar emits a beam of radiation. Every time the pulsar spins, this beam sweeps past Earth and astronomers detect a pulse of radiation. The gravity of a planet as it orbits the star pulls the pulsar slightly toward it and away from Earth. When the pulsar moves toward Earth, its pulses bunch up and more arrive per second than usual; when the pulsar moves away from Earth, the opposite happens, and fewer pulses arrive per second. By monitoring the pulses arriving from PSR B1257+12, Wolszczan and Frail discovered its three planets.

The planets circling this pulsar have masses comparable to that of Earth. In fact, the planets bear a surprising resemblance to Mercury, Venus and Earth, the three nearest planets to our own Sun. In our solar system, Mercury is the least massive of the three and the next two planets, Venus and Earth, have greater masses similar to each other. The same is true of the planets that orbit around PSR B1257+12.

Despite their resemblance to our solar system, pulsar planets probably can't support life because their stars emit deadly radiation. The best bets for life-bearing planets are those that circle Sun-like stars. In 1995, Swiss astronomers Michel Mayor and Didier Queloz made the headlines when they discovered the first extrasolar planet around a Sun-like star. The star, named 51 Pegasi, lies in the constellation Pegasus and is 50 light years from Earth. Mayor and Queloz discovered the planet because it, too, made its star wobble.

But this particular planet was completely unexpected. It is a giant planet like Jupiter and Saturn, yet it lies very close to its star. In fact, the planet lies even closer to its star than Mercury does to the Sun. Astronomers suspect that 51 Pegasi's planet formed far from its star and then spiralled inward to its present location.

Since the discovery of 51 Pegasi's planet, astronomers have discovered additional planets around Sun-like stars. Many of these are 'Pegasian': giant planets that orbit very close to their stars. But one Sun-like star has a solar system that mirrors our own. This star, known as 47 Ursae Majoris, has a giant planet about twice as far from its star as Earth is from the Sun. If this planet were a member of our solar system, it would lie between the orbits of Mars and Jupiter, but closer to Mars than to Jupiter.

Of course, many more discoveries have yet to be made, and some day astronomers will catalogue hundreds of extrasolar planets. Then they can at last compare other solar systems with our own to see whether our system is typical or atypical. Perhaps in 20 or 30 years, planet-hunters will achieve one of their ultimate goals: spotting planets the size of Earth circling stars like the Sun. Earth-sized planets around a star like the Sun could support life.

Once Earth-sized planets are seen directly, their spectra should reveal information about their compositions and structure. Some of these planets will be nearly airless worlds similar to Mercury, and will therefore be unlikely to support life. Others will show strong carbon dioxide signatures, as do Venus and Mars. But it is likely that a few planets will also show water vapour, indicating oceans of liquid water. And a few of those might even show atmospheric oxygen, a sign of life. Carbon-based life-forms were what pumped oxygen into Earth's atmosphere billions of years ago; if all life on Earth suddenly perished, the oxygen now present in Earth's atmosphere would react with surface rocks and disappear completely. Thus, if astronomers can detect large amounts of oxygen in another planet's atmosphere, that planet might well be supporting life.

Obviously such discoveries are decades away. But the exciting finds of the 1990s – of the first genuine planets circling other stars – have set the stage for a forthcoming century of a spectacular quest for planets. 99

A pulsar is a rapidly spinning neutron star, a star at the end of its life, that emits deadly radiation. This artist's impression shows a planet orbiting a pulsar. Three planets may orbit around the pulsar PSR B1257+12.

STARS

Stars appear to us as twinkling points of light in the night sky. They are, in fact, huge objects glowing with the heat from their inner nuclear furnace, which is so hot that it can change one element into another. All light, heat and energy on Earth comes from the nuclear furnace of our nearest star: the Sun. The Sun is a rather ordinary star, but looking outward we can see gigantic cool stars, red giants, bright, hot supergiants and small stars made of incredibly dense material, white dwarfs. There are also neutron stars and black holes, which are the dead remnants of massive stars. Many stars orbit around each other and, at times, are so close together that material from one is pulled off by the other. Some pulsate so much that astronomers can see their light varying from hundreds of light years away. Overall, the twinkling stars are strange and exotic objects, which allow astronomers to see way beyond our own solar system.

2 ► *Once a cloud starts to collapse, it fragments into small clumps because within it are small knots of gas and dust. The knots are denser than the cloud, and have a greater gravitational pull, so they gather more gas and dust into themselves.*

1 ► *Stars are formed from huge clouds of cold dust and gas. Dark clouds in the interstellar medium can be seen silhouetted against a brighter background. Some outside event is needed to disturb the cloud so that it starts to collapse in on itself.*

12 ► *Some initially very massive stars can collapse even further than the neutron stage to become black holes. These have such high gravity that they pull everything nearby into themselves.*

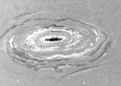

11 ► *Neutron stars are small, dim and dense, and can spin rapidly – up to many thousands of times a second. Their strong magnetic fields funnel radiation out, so beams of radiation sweep across the sky like a lighthouse. If the beams point at us, we hear rapid, regular pulses of radio waves. These stars are known as pulsars.*

10 ► *If a larger star collapses under its own gravity to become a neutron star, its outer envelope can be blown off in a massive explosion that is known as a supernova.*

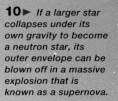

3 ▶ *As the knot grows, the pressure and temperature at the centre rise. When the temperature reaches about 10,000,000K, nuclear reactions will begin and a star is born.*

4 ▶ *The cloud of dust and gas spins around the shrinking central star. The spinning stops the cloud collapsing inward, so a flattened disk is formed.*

5 ▶ *As the disk of dust and gas cools, the material within it begins to clump together. The young star can react quite violently, and produce a very strong stellar wind. If the clumps of material are large and dense enough to avoid being blown away by this wind, they may become planets.*

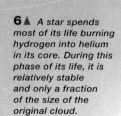

6 ▲ *A star spends most of its life burning hydrogen into helium in its core. During this phase of its life, it is relatively stable and only a fraction of the size of the original cloud.*

9 ◀ *Once all the nuclear fuel of a star has been used up, it will start to collapse. Small stars (up to nearly one-and-a-half times the mass of the Sun) end their lives as white dwarfs.*

7 ▲ *When a star has burned up all the hydrogen in its core, it starts to burn that in its atmosphere. As it does so, the star expands and cools.*

THE LIFE OF A STAR

All single stars undergo the same processes of evolution, although their size dictates how they will eventually end their lives. This diagram shows the evolution of a star that is similar in size to our Sun, together with the different later stages of the life of a giant star.

8 ◀ *Stars of about the Sun's mass become red giants; more massive stars become red supergiants like Betelgeuse in Orion.*

LIFECYCLE OF A STAR

A STAR CAN LIVE FOR BILLIONS OF YEARS AS IT DEVELOPS FROM A CLOUD OF DUST UNTIL IT REACHES WHAT CAN BE A DRAMATIC STELLAR DEATH.

Astar's lifecycle depends on its mass. A star with a very low mass, like our Sun, will have a calm life with a moderate end. However, a high-mass star, such as Betelgeuse, will have an extremely dramatic demise.

STELLAR LUMINOSITY

One obvious fact that can be seen when just looking into the night sky is that some stars are faint while others are very bright. One reason for this is the varying distances of the stars from the Earth. If you have two objects of similar brightness, then obviously the object that is the farthest from you will appear the dimmest.

This makes it difficult for astronomers to measure the brightness of stars. If the distance to a star can be found, then scientists can determine its absolute brightness as opposed to its apparent brightness in the sky.

The brightness of a star is related to its luminosity, which is the amount of energy emitted by the star every second. The Sun's luminosity is 3.9×10^{26} watts, equivalent to a million, million, million, million 100-watt light bulbs.

The absolute luminosity of a star, ie, its true brightness and not its brightness as it appears to us, varies with temperature and size. The hotter a body, the more energy it will radiate and the higher luminosity it will have. The larger the body, the more surface area there is for energy to radiate from, just as a large radiator will give off more heat than a small one at the same temperature. Therefore, a large cool star can have the same luminosity as a small hot star.

The luminosity of stars is vital in determining stellar lifecycles and is plotted on the vertical axis of the Hertzsprung-Russell diagram (below right).

THE HERTZSPRUNG-RUSSELL DIAGRAM

The Hertzsprung-Russell diagram is one of the most powerful diagrams in astronomy on which the whole of the life history of a star can be drawn.

It came about when astronomers were looking at the stars and attempting to formulate a pattern in what they were seeing. It is obvious, from looking with the naked eye only, that stars have different colours. They range from brilliant blue-white, through yellow to a dull red. And astronomers now know that these colours represent temperature, just as a rod of iron will glow red then white as it gets hotter in a fire.

Ejnar Hertzsprung (in 1905) plotted the absolute magnitude of stars against their colour index (how bright or dim a star is in different bands of light), and found a pattern.

A close-up taken by the Hubble Space Telescope in 1995 of part of the Eagle nebula. The finger-like protrusions seen at the top are larger than our own solar system. Inside each protrusion is an embryo star.

◀ 16–19 Formation of the solar system

◀ 95 Atomic spectra

This same pattern was discovered by Henry Russell 10 years later when he plotted absolute magnitude against spectral type.

The spectral type is based on the spectra of stars, which show absorption and emission lines. The types fall into broad classes from the hot bright O stars, through classes B, A, F, G, K, down to the cool red M stars. Each class is further divided into 10 from 0 to 9. The spectral class of the Sun is G2.

HOW TO USE THE H-R DIAGRAM

At the time Hertzsprung and Russell plotted their diagrams, they did not realize that the colour index and spectral type of a star are measurements of a star's temperature. Russell plotted spectral types along the bottom axis from O to M. Astronomers now know that this classification indicates a progression from hot to cool stars, so temperature on the Hertzsprung-Russell diagram (or H-R diagram) goes backward along the bottom axis. On the vertical axis, absolute magnitude indicates the star's luminosity.

The wavy band indicates the lifecycle of a main sequence star (see below), which travels from the bottom right corner to the top left. The Sun is currently on this wavy band at a temperature of about 6000K. Groups in the top right-hand part of the graph represent giants and supergiants and white dwarfs appear at the bottom.

THE MAIN SEQUENCE

For most of its life, a star will fuse hydrogen in its core, creating helium, and it is known as a main sequence star.

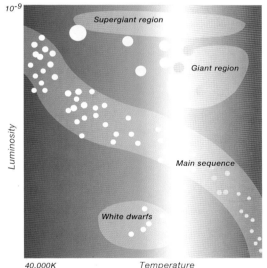

10^{-9}

Supergiant region

Giant region

Main sequence

White dwarfs

Luminosity

40,000K   Temperature

HERTZSPRUNG-RUSSELL DIAGRAM

Above: The dark area in this image is a star-forming cloud of dust and gas, which is obscuring the light from objects farther away. The group of bluish stars on the left are young stars born of such a cloud.

This nuclear fusion, changing one element to another, is brought about by the immense pressure and temperature at the centre of a star. Since stars spend most of their lives in this stage, most of the stars you can see in the sky are on the main sequence and are relatively stable.

THE BIRTH OF A GIANT

The next stage of a star's life depends on its mass. The Sun has been burning hydrogen in its core for around 4.6 billion years, and will continue to do so for another five billion years. Stars that are more massive than the Sun will

Spectroscopy 142 ▶

Stellar magnitudes 166–167 ▶

In 1995, the Hubble Space Telescope revealed a number of white dwarf stars (circled) in the globular cluster M4. These incredibly dense old stars gradually cool and fade until they can no longer be seen.

The Hourglass nebula is a planetary nebula surrounding a central star imaged by the Hubble Space Telescope in 1996. The hourglass shape is thought to be caused by the stellar wind reacting with associated magnetic fields.

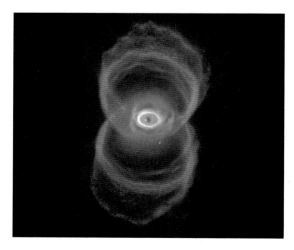

burn through their store of hydrogen faster and will not remain on the main sequence for as long. Stars that are less massive may last longer in this state.

When a star has finished fusing all the hydrogen in its core, it will burn the hydrogen in its atmosphere, or envelope, which it burns in thin spherical shells, known as shell hydrogen burning. The hydrogen burning shells gradually work their way outward, causing the outer layers to expand, and as the star's outer atmosphere expands, it cools.

A low-mass star, such as our Sun, will become a red giant at this stage. When it has finished its core hydrogen burning, it will cool to a temperature of about 3500K and will swell out to a diameter of about 1AU, gobbling up the planets Mercury, Venus and Earth.

More massive stars, such as Betelgeuse, become larger than red giants; they become supergiants. The diameter of Betelgeuse is about 500 times that of the Sun.

RUNNING OUT OF FUEL

When red giants and supergiants have used up all the hydrogen, their cores, which will be made up primarily of helium as a result of hydrogen burning, will collapse. Depending on how massive the star is, this core can collapse to pressures and temperatures high enough to start the nuclear burning of this helium.

The star behaves in a manner similar to that when it was burning hydrogen, but it will not take as long to use up all the helium fuel. A red giant will probably spend only two billion years burning helium in its core. Then helium shell burning occurs, and the core once again collapses.

The Sun will do no further nuclear burning, but more massive stars can go on to use up all the helium. In very massive supergiants, nuclear fusion reactions can occur until iron is produced, which is the heaviest metal that can be formed by this method.

STELLAR DEATH: WHITE DWARFS

In a low-mass star, once all the nuclear fuel has been used up in its core, the core will collapse until electron degeneracy sets in. This is when the force between electrons is sufficiently great to overcome any further core collapse.

This event happens only when the pressure is immense, and the core material is thus much denser than usual stellar material. A star that reaches this stage is called a white dwarf.

White dwarfs are small and very dense. A matchbox full of white dwarf material would weigh about 15,000kg, about as heavy as 15 cars. White dwarfs eventually fade from view as they cool.

NEUTRON STARS AND PULSARS

If the initial star was massive enough, the weight of infalling material will be enough to overcome the electron degeneracy pressure and the core will continue to collapse. Then neutron degeneracy pressure can set in: the force between individual neutrons, which will stop the core contracting.

A star that reaches this stage is called a neutron star. A matchbox full of neutron star material would weigh about 15×10^{11}kg, about as heavy as one and a half million million cars.

Rapidly rotating neutron stars with intense magnetic fields are the sources behind pulsars. If correctly oriented, beams of radiation from pulsars sweep across the Earth like a lighthouse beacon. A brief flash of radiation is seen each time the beam sweeps across. Some pulsars are rotating so rapidly that the pulses of radiation are seen every few milliseconds. This means the neutron star must be rotating hundreds of times each second.

SUPERNOVAE

When neutron degeneracy pressure sets in, it can sometimes do so very rapidly, presenting a rigid surface to the still infalling material. At this point scientists believe a supernova occurs; with the shock of meeting this rigid surface, the remaining outer layers of the star are blown off in an enormous explosion. These explosions can be so massive that a supernova can outshine an entire galaxy.

Over the following weeks and months the spectacle dims, but the debris from a supernova can remain visible for tens of thousands of years afterwards. This debris is enriched in some of the heavier chemical elements as a result of the nuclear reactions that took place in the star and in the supernova explosion. After the debris has scattered, a spinning neutron star can be left.

Supernovae are the only places in the universe where gold can be formed. Typically, only a few supernovae occur in a galaxy in a century, which is why gold is rare.

BLACK HOLES

Some stars are massive enough to have sufficient infalling material to overcome even neutron degeneracy pressure. There is, therefore, nothing further to stop its collapse, and it will continue to collapse in on itself, becoming denser and denser until it forms a black hole.

Black holes are so dense and have such strong gravitational fields that not even light can escape.

THE SOLITARY SUN

Most of stellar evolutionary theory is based upon a solitary star, such as our Sun. However, most stars are members of a binary or multiple star system. Over 60 per cent of stars appear to be members of such a system, and this can have a dramatic effect on their evolution.

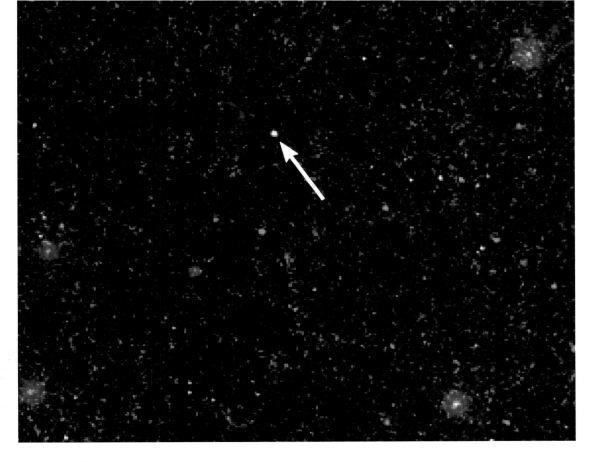

Above left and right: These images show the before and after effects of a supernova explosion. This supernova took place in 1987 in the nearby galaxy, the Large Magellanic cloud. The arrow on the left-hand image shows the size of the star before the dramatic explosion in the right-hand image.

Left: The Hubble Space Telescope found an isolated neutron star in 1997. It is no more than 28km across and its surface temperature is more than 666,000°C.

Double stars 104–106 ▶

Betelgeuse 168 ▶

DOUBLES AND VARIABLES

OUR SUN IS NOT REPRESENTATIVE OF ALL STARS.
SOME ARE PART OF DOUBLE SYSTEMS AND SOME
VARY NOTICEABLY IN BRIGHTNESS OVER TIME.

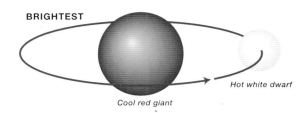

BRIGHTEST

Cool red giant *Hot white dwarf*

EDOUARD ROCHE
Edouard Roche was a French mathematician who worked during the mid-1800s. One of his main discoveries was concerned with the forces of gravity of binary stars. He was the first to realize that if a 'figure of eight' were to be drawn around two stars in a binary system, then this figure would define the gravitational domain of each of the stars. Each half of the figure of eight is known as a Roche lobe. If either star swells enough to fill its Roche lobe, then mass will be transferred to the other star through the point where the two Roche lobes touch. This point is called the inner Lagrangian point.

B inaries are pairs of stars that orbit around each other. The two stars form near to each other and eventually become bound by gravity. Traditionally, astronomers divide binaries into three classes: visual, spectroscopic and eclipsing. Stars can also belong to multiple systems where more than two stars orbit around a common centre of gravity. Multiple systems are usually hierarchies of the same scenario, a distant star going around a double, or two doubles orbiting each other.

VISUAL BINARIES

Visual binaries are pairs of stars that can be separated through a telescope: good examples are Mizar (below right and opposite) in the Plough (the Big Dipper), gamma Andromedae and gamma Leonis.

In this type of binary star, the components are so far apart that they take many, even thousands, of years to orbit each other. Over time, astronomers can measure the position of the fainter star relative to the brighter and can construct their elliptical orbits.

It is then possible to work out the stars' masses using Kepler's third law. Two bodies always orbit a common centre of mass placed between them whose location relative to the centres of the stars is inversely proportional to the masses of the components. Location of the centre of mass will give the mass ratio and, with the sum of the masses, astronomers find the individual masses of the stars in terms of solar mass.

It is normally difficult to observe the movement of binary stars because we will not often see the orbital plane face-on. Imagine you are watching a man walking around a stationary object. You will not be able to see his movement as he is moving away from you or toward you but will be able to see him walking from side to side. This process can

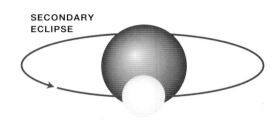

ECLIPSE

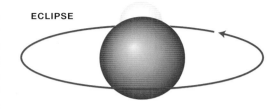

SECONDARY ECLIPSE

ECLIPSING BINARY STARS
Some binary stars are orbiting in such a way that they pass in front of one another and eclipse each other. This causes variations in brightness often visible from Earth.

be complicated as we see the binary stars at different angles in the sky, but if both stars are seen, it is possible to work out the orbit taking the inclination into account.

SPECTROSCOPIC BINARIES

Spectroscopic binaries are pairs of stars that are so close together that they can only be resolved by spectroscopy: the study of the absorption and emission lines in the spectra of the stars.

Spectroscopic binaries are much closer together than visual binaries and therefore they orbit each other at a faster rate. These high velocities will produce easily observable Doppler shifts in the combined spectra of the two stars.

Alpha and beta Centauri (right) form a bright naked-eye pair a little over four light years from us, orbiting each other every 80 years. Mizar (far right), in Ursa Major, is 59 light years away and observers need a small telescope to separate the two stars, which take 20,000 years to orbit each other.

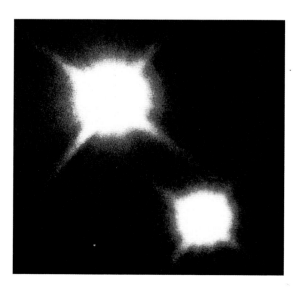

If the plane of the orbit is in the line of sight, astronomers can measure orbital speeds. The orbital period combined with the speed gives the size of each orbit around the common centre of mass, and thus the individual masses. However, the orbit will more often be tilted, which means that scientists will only be able to calculate the lower limits of the speeds and masses of the stars.

Also, if one star is much fainter than the other, astronomers will only be able to see one spectrum, which further limits what can be learned. However, if we can guess the mass of the brighter star from its spectrum, we can derive a lower limit to the mass of a low-mass companion. This method has enabled astronomers to discover Jupiter-sized planets.

Advances in technology are such that astronomers are able to observe more and more visual binaries: interferometers are now able to separate stars that are only a few thousandths of an arc second apart. Therefore, the methods overlap and complement each other to a certain extent. Once binaries have been discovered using spectroscopic techniques, they can then be probed with interferometers and eventually be resolved visually.

However, only very bright stars can be observed interferometrically, so spectroscopy remains a very important tool in binary analysis.

ECLIPSING BINARIES

Eclipsing binaries are aligned in such a way that each of the stars can get in the way of the other and therefore they can eclipse each other. From the Earth, astronomers can watch the binary system dim during an orbital period, as one star passes in front of the other.

Several eclipsing stars are visible to the naked eye, the most famous of which is Algol in Perseus, which drops by as much as 1.5 magnitudes every 2.8 days as one star covers, but does not completely hide, the other.

The depths of eclipse depend on the sizes of the stars and on the relative brightnesses of companions, for which there are numerous possibilities. A spectacular eclipse takes place when a small, hot, bright dwarf orbits a large cool giant. When the dwarf disappears behind the giant, the combined visual brightness can drop one or more magnitudes. When the bright dwarf is in front, however, it cuts off only a part of the giant's light, leading to a shallow 'secondary' eclipse (see opposite top).

Astronomers can learn an enormous amount by observing eclipsing stars. For there to be a good probability of eclipse, the stars must be close together, making them spectroscopically observable. The shape of the 'light curve', ie, the variations in brightness as the stars pass in front of and behind each other, allows astronomers to find out the tilt of the orbit, the speed at which the stars are moving and finally to calculate the mass of the two stars.

Once astronomers know the speed at which the stars are moving, the durations of the eclipses enable them to work out the stellar diameters. Once they have calculated the stars' distances and brightness,

Mizar and Alcor are a wide visual binary system in the handle of the Plough or Big Dipper – the second bright star from the left of this image. They look as if they are close together from Earth but are at least a light year apart.

Mira or Mira Ceti is a variable red giant in the constellation Cygnus and is circled in the images. It is the prototype of all Mira-type stars, which are cool red giants varying in brightness because of pulsations.

they can then work out the surface temperatures of the stars. The shape of the light curve even enables them to distinguish between the dark edges and bright centres of the stars.

The results of all these studies show that main sequence brightnesses of stars are roughly proportional to the cube of their mass; a fundamental relationship that is well explained by scientifc theory.

VARIABLE STARS

The eclipsing binary is a type of 'variable star', a star that changes its brightness with time. The eclipsing star is variable because of the geometry of its orbit. However, a huge number of single variable stars are also known to exist: these include Mira-type, pulsating, Cepheids and RR Lyrae, explosive, cataclysmic and symbiotic.

MIRA-TYPE VARIABLES

Astronomers have identified dozens of types of variable stars. The most prominent is named after the star Mira (the 'amazing one') in Cetus, known since 1596. Mira is sometimes among the brightest stars of its constellation, usually reaching third magnitude, sometimes even first. Most of the time, however, it is invisible to the naked eye. Its variation period of 332 days leads it to be called a 'long period variable'.

Thousands of Mira-type stars are known. All are cool red giants (classes M5 to M9). Most have chemical compositions similar to that of the Sun, with more oxygen than carbon, but a significant number are 'carbon stars', with more carbon than oxygen: the result of convection currents dredging up the by-products of helium burning.

The variation in brightness can be seven or more magnitudes, with periods ranging from about 100 days to over 1000. The longer-period stars are generally the most luminous. Much of the visual brightness is caused by small variations in surface temperature: lower temperatures move the radiation into the infrared part of

the electromagnetic spectrum, and the variation in the infrared is considerably smaller.

PULSATING STARS

Some stars vary in brightness because of physical pulsations, which change their size by alternately swelling and contracting. These pulsations are caused when deep layers alternately trap and release radiation during the ionization of hydrogen or helium.

Some variable stars are in the process of losing their entire envelopes back into space, revealing their cores. The lost mass will be lit by the hot cores and turn into lovely but fleeting planetary nebulae. The outbound matter enriches interstellar space with by-products of nuclear reactions, such as carbon. Most of the dust in the galaxy is thought to come from such stars. The stellar cores eventually make the white dwarfs that swarm around us.

The process by which stars lose thier outer envelopes is caused by shock waves in the stellar atmospheres, which lift gas from the surface. In the cooler outer reaches, the escaping gas chills and partially condenses into dust: oxygen-rich stars making silicate dust, carbon-rich stars carbon dust. Pressure from stellar radiation blows the dust outward, and the dust grains drag the gas to create a powerful wind that, in extreme cases, can reach a mass loss-rate of $1/100,000$ of a solar mass a year. Some stars even disappear completely within the dust shrouds.

ROCHE LOBES

Surrounding any binary star is a figure-of-eight-shaped surface, the Roche lobe. This is where gravitational energy is constant as a result of the combined gravity of the stars and of centrifugal force. At the contact point directly between the stars, the gravitational force is effectively zero. If one star expands to fill its lobe and make contact with that point, mass can flow from that star to its companion. For this to happen, the stars must be close enough together that ordinary single star evolution causes sufficient expansion to allow one of the stars to fill its lobe.

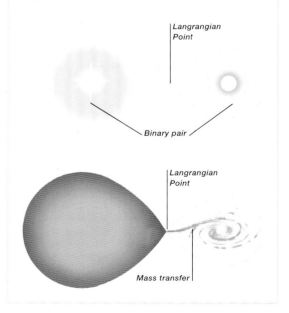

Langrangian Point

Binary pair

Langrangian Point

Mass transfer

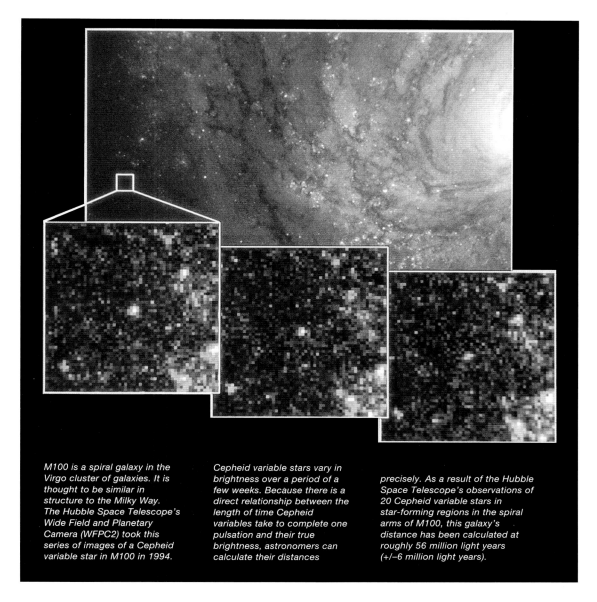

M100 is a spiral galaxy in the Virgo cluster of galaxies. It is thought to be similar in structure to the Milky Way. The Hubble Space Telescope's Wide Field and Planetary Camera (WFPC2) took this series of images of a Cepheid variable star in M100 in 1994.

Cepheid variable stars vary in brightness over a period of a few weeks. Because there is a direct relationship between the length of time Cepheid variables take to complete one pulsation and their true brightness, astronomers can calculate their distances

precisely. As a result of the Hubble Space Telescope's observations of 20 Cepheid variable stars in star-forming regions in the spiral arms of M100, this galaxy's distance has been calculated at roughly 56 million light years (+/–6 million light years).

CEPHEIDS AND RR LYRAE STARS

Another type of variable star that is as common as Mira-type stars is Cepheids. They were named after the first one that was discovered, delta Cephei. It varies by approximately one stellar magnitude over a 5.4-day interval. These stars are high-mass unstable supergiants of spectral type F and G that are moving through an 'instability strip' that runs down the middle of the Hertzsprung-Russell diagram.

Their variation periods range from as little as one day to over 100 days and their brightness sometimes varies only subtly and sometimes by as much as two magnitudes or more.

Cepheids are mainly found in the galaxy's disk. The galaxy's halo contains short-period versions of Cepheids, the RR Lyrae stars, which vary less than a magnitude in brightness and in periods of less than a day.

In addition, there are semiregular and irregular giants, supergiants and white dwarfs, the latter changing every few minutes (ZZ Ceti stars), and stars that oscillate at varying intervals (beta Canis Majoris stars).

MASS TRANSFER

Binaries can be variable because of the geometry of their orbit, single stars as a result of pulsation. But binary systems can produce other kinds of variations as a result of the interaction between the two stars and these results can be spectacular.

Of the two components of a binary, the more massive star will evolve first, becoming a giant and then a white dwarf. While in the giant state it may fill a tidal surface called a Roche lobe in which the force of gravity is effectively zero. Some mass flows to the smaller companion, but if the rate of mass loss is great enough, some may spill outward to encompass the pair in a common envelope.

Friction then causes the two stars to spiral closer together. The giant turns into a white dwarf and, if the binary components are close enough, the fun begins.

CATACLYSMIC VARIABLES

Cataclysmic variables include novae and recurrent novae. A nova is a spectacular event where a binary

The light from the Nova Cygni, a symbiotic star, reached the Earth in 1975. The system is now surrounded by an expanding, asymmetric shell of gas that was ejected by the explosion.

star system brightens suddenly by as much as 12 magnitudes. From the Earth we see a 'nova' erupt into the night sky. A bright naked-eye nova takes place about once a generation, but several can be seen through telescopes each year.

Novae are brought about by a thermonuclear reaction, which takes place on the white dwarf. Tides raised by the white dwarf in its ordinary dwarf companion may cause the small hot star to fill its Roche lobe and spill hydrogen-rich matter toward the white dwarf companion. The mass will flow first into a disk surrounding the white dwarf, from which fresh hydrogen falls onto its surface.

When the temperature and pressure at the base of this hydrogen layer become great enough, a thermonuclear explosion blasts away the surface of the white dwarf causing the increase in brightness.

The transfer of mass between the two stars resumes following the outburst and, after a few hundreds of thousands of years, the nova probably repeats itself. Classical novae, by definition, have never been seen to repeat, but there is a version called a recurrent nova that does repeat with less power over intervals of just a few decades.

Between explosions the stars will almost certainly be unstable, and will flicker and produce minor outbursts, creating 'dwarf novae'.

SYMBIOTIC STARS

If the stars are farther apart, the interaction does not begin until the dwarf swells up into a giant that then sends considerable mass to the white dwarf. The hot accretion disk produces enough ultraviolet radiation to cause the flowing matter to fluoresce and generate emission lines. We now see a symbiotic star, which has the characteristics of a cool giant coupled with features caused by high temperatures.

Symbiotic stars sometimes also become unstable, producing outbursts. One, CH Cygni (above), visible to the naked eye, has been in an outburst state for years.

If the giant is a Mira-type star, the variations become even more complex, as the single- and double-star variations are combined.

OTHER DRAMATIC COMBINATIONS

In the ultimate binary pairing the white dwarf is so massive that it is close to becoming a neutron star. The accreting matter might therefore push the white dwarf over the limit before a nova can take place. The white dwarf then collapses catastrophically. The resultant nuclear explosion causes a supernova that is even brighter than the iron-core collapse variety seen among the massive single stars.

Astronomers need these types of scenarios to explain the supernovae that occur in places where there are no

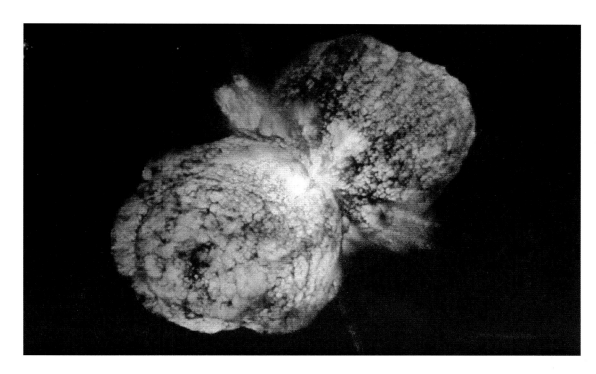

Eta Carinae is a massive luminous blue variable. In the 19th century it ejected this huge cloud of gas and dust. The star seems destined to explode as a core-collapse supernova.

HENRIETTA LEAVITT
Henrietta Leavitt (1868–1921) was head of photographic photometry at Harvard College Observatory. She specialized in photographic analysis of variable star brightness and is best known for her discovery of the relationship between period and luminosity of Cepheid variable stars. This breakthrough discovery formed a basis for the method used by scientists to determine the distances of stars.

such massive stars, such as the halos of galaxies and elliptical galaxies. Double white dwarfs in which the components merge might also be responsible for some supernovae explosions.

In 1572, Tycho Brahe recorded a brilliant star in the constellation Cassiopeia that rivalled Venus in apparent brightness (see below). Reconstruction of its light curve as well as modern observation of the remnant gas cloud strongly suggest that it was a white dwarf type of supernova, as was the great nearby supernova of 1006 in Lupus, attested to by a Chinese scribe with the words 'things could be seen by its light'.

Other events, such as unusual X-ray outbursts and novae, are observed when mass is transferred into a neutron star or a black hole.

MEASURING DISTANCES

Astronomers need to know the distances to stars to enable them to determine other characteristics, for example how luminous they are. However, because of the huge distances involved, it is not an easy task.

The method of parallax has been used by astronomers to determine the distances to over one hundred thousand of the nearest stars to the Earth. As our planet travels in its orbit round the Sun, the nearest stars 'move' slightly against the background of stars farther away. This movement is due to the Earth's motion. By measuring the angle the star has apparently moved, and knowing the distance of the Earth from the Sun, the distance to the star can be determined.

Astronomers have also needed to find the distances to other objects, not just stars. In 1912, Henrietta Leavitt carried out a study of the cycles of about 25 Cepheid variables in the Magellanic clouds (nearby companion galaxies). Her results showed that the luminosities of the Cepheids correlated strictly with their pulsation periods. To find the distance

of a Cepheid, astronomers need only measure its variation period, which immediately gives the star's average absolute magnitude. Knowing the average apparent magnitude (and allowing for the dimming effect of interstellar dust) yields the distance.

Cepheids are so bright that they can easily be seen in other galaxies. It was Edwin Hubble's identification of Cepheids in the Andromeda galaxy, M31, that allowed him to measure its great distance and determine that it really is an external galaxy.

The white dwarf supernovae provide an equally important means of measuring distance. Astronomers have discovered that the maximum brightnesses of such supernovae are almost equal. By examining the explosion, they can determine the maximum apparent magnitude and by comparing this value with the absolute magnitude, they can find the distance. Supernovae are so bright that they can be seen over great distances, much farther than Cepheids, and this enables astronomers to find the distance to much more distant galaxies and to measure the Hubble constant and the age of the universe.

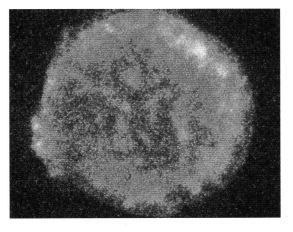

This image is taken in X-ray and shows the shockwave of the supernova that Tycho Brahe saw explode dramatically in 1572.

GALAXIES

From a dark site, a meandering glowing band can be seen stretching across the night sky. This is the Milky Way, the light from thousands of individual stars lying in the plane of our own galaxy. Within our galaxy are a thousand million stars, and from our galaxy we can see thousands of other galaxies. The Milky Way galaxy is a spiral galaxy and the Sun lies on one of the spiral arms. Close to the Milky Way, visible from the southern hemisphere, lie the Magellanic clouds: two irregular patches of light, which are, in fact, companion galaxies to our own. Some galaxies are spirals like ours, some are elliptical, but most galaxies are irregular. Often they are irregular because of dramatic events occurring within them. In looking at faraway galaxies, astronomers are looking back in time. Studying galaxies can reveal secrets about the history of the universe.

SIDE VIEW
Seen edge-on, the central bulge of the Milky Way is very obvious. Star formation takes place in the plane of the galaxy, while the halo of globular clusters surrounding the galaxy is home to older stars.

1 Galactic plane of dust and gas
2 Position of the Sun
3 Galactic nucleus
4 Globular clusters in galactic halo

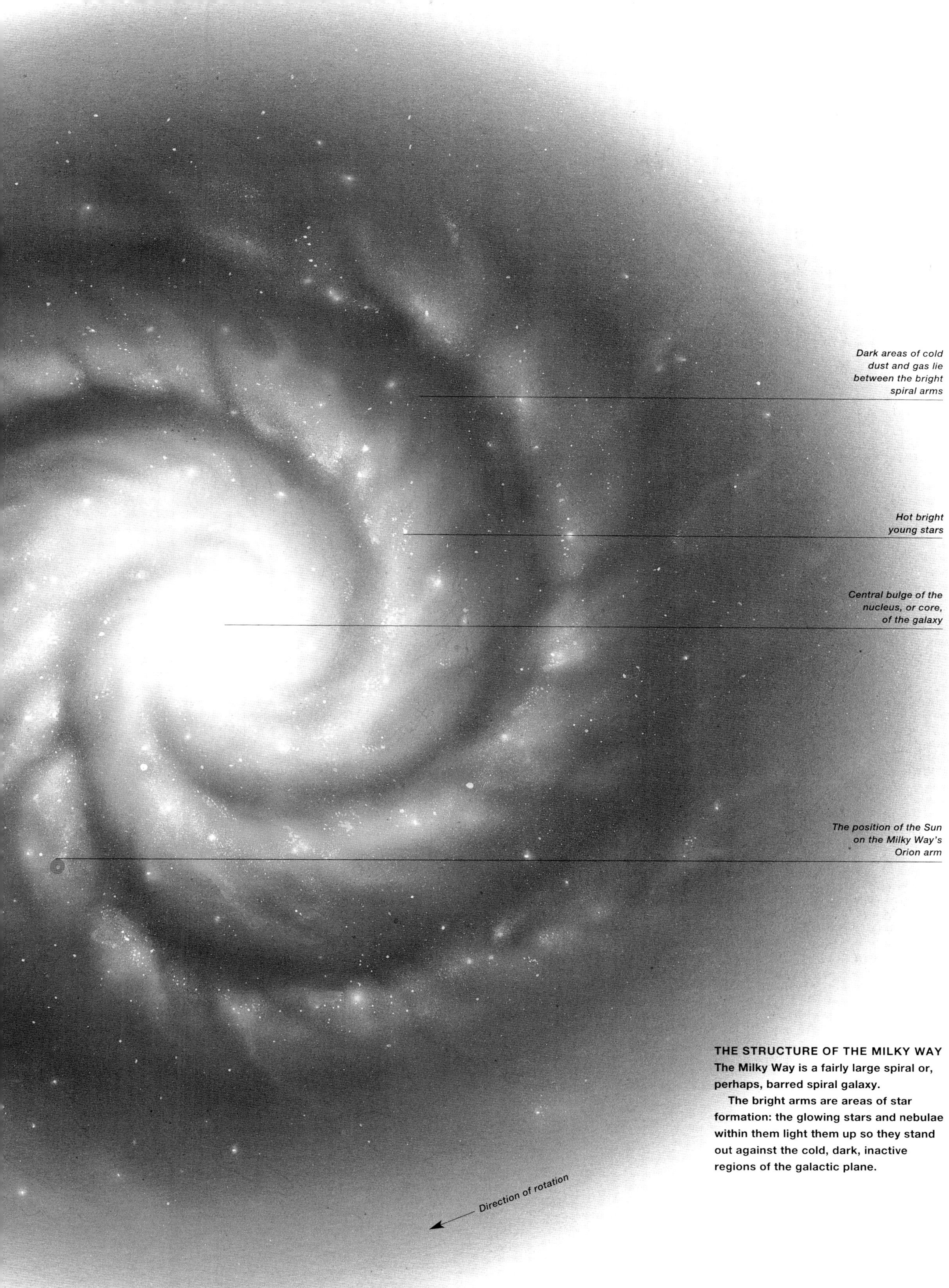

Dark areas of cold
dust and gas lie
between the bright
spiral arms

Hot bright
young stars

Central bulge of the
nucleus, or core,
of the galaxy

The position of the Sun
on the Milky Way's
Orion arm

Direction of rotation

THE STRUCTURE OF THE MILKY WAY
The Milky Way is a fairly large spiral or,
perhaps, barred spiral galaxy.
    The bright arms are areas of star
formation: the glowing stars and nebulae
within them light them up so they stand
out against the cold, dark, inactive
regions of the galactic plane.

THE MILKY WAY

THE MILKY WAY IS SO NAMED BECAUSE IT APPEARS IN THE SKY AS A BROAD MISTY BAND OF STARS THAT STRETCHES FROM HORIZON TO HORIZON.

Our galaxy is all around us and is impossible to capture in one photograph. This composite image was made from many photographs and shows the whole 360° span of the Milky Way.

The arc of the Milky Way is a remarkable sight, especially in the skies of the southern winter. Astronomers now know this band of light is our galaxy, a huge, self-contained spiral star system viewed from within. It was not until this century that astronomers understood what type of galaxy we live in and its structure. The Milky Way is a spiral galaxy, with a bright nucleus and bulge at the centre, surrounded by orbiting spiral arms of stars, nebulae and dust. It appears broadest and brightest in the southern constellation of Sagittarius, but almost everywhere seems to be divided in two by dark patches of dust of variable size and blackness.

EARLY IDEAS

The discovery that the Milky Way is just one of billions of galaxies, each of which contains billions of stars, has come only gradually, and our understanding is still far from complete. The first inklings came in 1610 when Galileo's telescope revealed that the Milky Way is made up of vast numbers of faint stars. The first scientific study was begun by Sir William Herschel who, in 1784, set himself the task of counting stars in more than 3000 selected regions of the sky. Because so little was known about stars at that time, he made a number of assumptions that led to the wrong conclusions. He assumed that all stars were equally luminous, so the fainter a star was, the farther away it must be. He imagined that he could see to the edges of the stellar system and, because he believed that stars were uniformly distributed, came to the conclusion that greater star numbers in an area of the sky indicated that the Milky Way

William Herschel's diagram of what he thought was the shape of the galaxy was the result of years of patient observation and star counting. Although the idea that the Milky Way is a flattened disk is correct, he had no way of knowing that he could see only a part of the system and that the Sun is not near the centre.

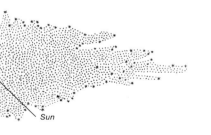

Sun

stretched farther in that direction. He could not know then that a large part of the galaxy was hidden from view. His diagram – the result of many years of work – represented a flattened disk with the Sun near the centre.

HERSCHEL AND MESSIER'S FUZZY BLOBS

Herschel had noted dark markings in the Milky Way, which he thought were signs of decay in its fabric. He also listed thousands of faint, fuzzy objects that did not seem to be made of stars, even when seen under the best conditions.

Charles Messier, a French comet hunter, also noted the misty patches while compiling the first systematic catalogue of non-stellar objects. Between 1764 and 1784 he listed about 100 extended non-stellar objects visible from Paris that looked like comets. These objects were fixed in the sky and were of no interest to him: he listed their positions so he could ignore them. Messier's catalogue is still used today because he noted some of the northern sky's most spectacular objects; many of these are known by their Messier number, such as the Andromeda galaxy, which is catalogued as M31.

Herschel suspected that many of these misty patches were in fact galaxies like the Milky Way, which were far away. And in 1845, William Parsons, Earl of Rosse, found that M51 had a spiral structure, and he assumed it was a rotating disk of stars. Like Herschel, Rosse had no way of knowing the distance or dimensions of his starry spiral because he had no way to analyse and record its feeble light. As so often happens in science, it was an advance in technology that provided the key to a long-standing puzzle.

A STAR AMONG STARS

The nature and composition of stars and nebulae was thought to be beyond understanding, particularly after stellar distances were first measured in 1838. In the 1860s, however, spectroscopes revealed dark lines in the Sun's rainbow of colours. These dark markings in the solar spectrum revealed the presence of vaporized elements in its atmosphere. Identical lines were seen in other stars, suggesting the possibility of analysing them and confirming that the Sun was one star among many. Only the most prominent lines of the brightest stars were visible, and it was not until detailed photography was used to record stellar spectra in the 1870s that astronomers could identify the characteristics of different types of stars.

Long photographic exposures were also used to produce images that revealed the structure of nebulous objects too faint to be studied by eye, and soon cameras began to detect nebulae that were too faint to be seen with even the largest telescopes. Many more 'spiral nebulae' were found, as well as others that seemed irregular or even amorphous.

Spectroscopy and photography began to show that there were different types of nebulae: some were made of vast numbers of stars too distant to be seen as individuals, others were clouds of glowing gas excited by hot stars in and around them. Some nebulae shined by reflected starlight revealing that they were made up of clouds of dusty particles.

A LARGE GALAXY

As the structure and composition of the components of the visible universe were gradually uncovered it was not clear if everything that could be seen was part of the Milky Way – implying that the Milky Way was the universe – or if there was something, anything, beyond it.

The main difficulty was the same one that Sir William Herschel had encountered – the astronomers' inability to measure the distances to the nebulae. The new science of astrophysics uncovered more and more facts about stars and nebulae, but at the beginning of the 20th century the Milky Way and its place in the universe remained a mystery.

In 1914, Harlow Shapley of Mount Wilson Observatory determined the distances to several globular clusters.

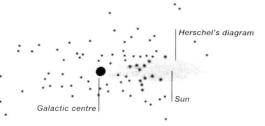

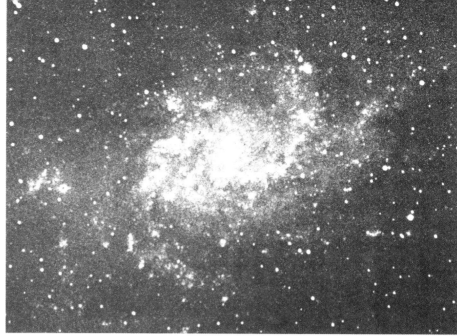

Harlow Shapley plotted the positions of many globular clusters and realized that they must orbit around the same centre as the Milky Way. Compared with William Herschel's diagram, it shows how much of the galaxy was hidden from the earlier astronomer's view.

These spherical clusters of stars seemed to be scattered widely across the sky – most of them in the southern Milky Way. Shapley measured the distances using a variable star technique in a similar manner to that used by Henrietta Leavitt to estimate the distance to the Small Magellanic cloud.

Although Shapley's distance measurements were interesting in themselves, he came to a far more dramatic conclusion, suggesting that the centre of the cloud of globular clusters in space was also the centre of our galaxy. He measured this position at about 50,000 light years from the Sun in the direction of Sagittarius and calculated that the Milky Way was about 300,000 light years across.

Other astronomers have since shown that Shapley's basic conclusion was correct and that the Sun is in the Milky Way's suburbs, not at the centre of the galaxy. However, Shapley's distances were much too generous, partly because the variable stars in globular clusters are not the same as those in the Small Magellanic cloud and partly because the effects of interstellar dust make the stars appear fainter and, therefore, more distant.

His gigantic dimensions for the Milky Way convinced Shapley that the spiral nebulae were a part of it and supported a general feeling that a Milky Way so huge must be at the focus of the universe. Not everyone agreed, and chief among the sceptics was Heber Curtis of Lick Observatory in California, who had long supported an idea known as the 'island universe' theory – believing that the Milky Way was only one among many spiral nebulae and was much smaller than Shapley supposed.

Astronomers now know that Curtis was right to assert that 'spiral nebulae' were comparable to our own galaxy, though his reasoning and distance scale were faulty. Shapley's scale was also wrong; he followed a sound line of argument, but it was based on incomplete and flawed data.

A GALAXY AMONG GALAXIES

Four years later in 1918 everything changed when American astronomer Edwin Hubble discovered that the nebulae M31

The Milky Way is believed to be similar to the spiral galaxy NGC2997 (below), although its star formation is thought to be more intense and the arms brighter than in our own galaxy.

The pioneering astrophotographer Isaac Roberts took this picture of the galaxy M33 in 1895. The nature of galaxies was not fully understood then; some astronomers believed M33 to be the result of stellar collisions.

THE MILKY WAY'S VITAL STATISTICS

- **Estimated mass: 1000 billion times the mass of the Sun**
- **Number of stars: 300–400 billion**
- **Estimated diameter of dark halo: 300,000 light years**
- **Diameter of visible disk: 100,000 light years**
- **Thickness of stellar disk (at Sun's location): about 1000 light years**
- **Diameter of the central bulge: 3000 light years**
- **Sun's distance from galactic centre: 30,000 light years**
- **Rotation rate (at Sun's location): 250km per second**
- **Orbital period (at Sun's location): 250 million years**
- **Luminosity: 20 billion solar luminosities**
- **Age: about 12 billion years**
- **Hubble type: uncertain, Sab to Sbc**

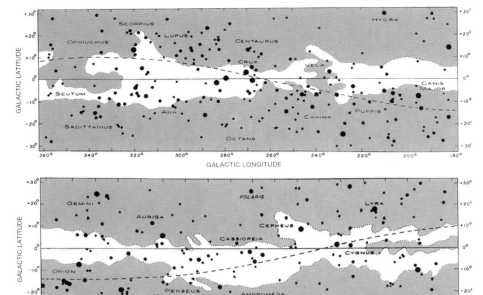

As astronomers look around the sky they can see a scattering of bright stars aligned with the diffuse Milky Way. This is Gould's belt, direct evidence of a young population of luminous stars in the galactic disk.

M16, the Eagle nebula, is an emission nebula in the constellation of Serpens. The dust and gas are glowing with the radiation from a cluster of young, hot, blue and white stars.

and M33 were in fact other galaxies like the Milky Way. He found that they were made up of millions of individual stars and, a year later, found Cepheid variables in M31, the Andromeda galaxy. These nebulae were much too distant to be part of the Milky Way, even within Shapley's enormous galaxy, and so Hubble determined that they must be galaxies in their own right. It was eventually discovered that there was more than one kind of Cepheid variable and that Hubble's measurements of M31 were incorrect, but that mattered little at the time. Hubble had established that the Milky Way was one galaxy among millions and that the universe was much more vast than anyone had imagined.

INTERSTELLAR DIMMING

With Hubble's momentous discovery, many observational inconsistencies now fell into place, and in 1930 another major cause of confusion and error was finally laid to rest when the Swiss-born American astronomer Robert Trumpler confirmed that the light of distant stars was dimmed by dust in space. Herschel's 'holes in the fabric of the galaxy' turned out to be dust, not decay.

STRUCTURE OF OUR GALAXY

The Milky Way is a spiral galaxy; it has a bright nucleus, a prominent central bulge around which the stars orbit in spiral arms. It is thin in comparison to its width and if viewed from side-on, it looks like a flattened disk. It is constantly revolving at 250km/sec but it takes the Sun about 225 million years to complete one circuit of the nucleus.

The Milky Way is the plane of our galaxy, within which the Sun and other stars orbit the nucleus as part of the spiral structure. It took so long for astronomers to make sense of the Milky Way because the Sun is buried in a disk of stars, gas and dust hundreds of light years thick, which makes up the spiral arms. The appearance of the Milky Way in the sky enabled astronomers to determine that our galaxy is a thin disk because fewer stars are seen at right angles to it.

The brightest naked-eye stars are concentrated towards the galactic plane in a band known as Gould's belt (see left), which follows the diffuse Milky Way. The belt is made of mostly hot, extremely luminous young stars, defined by astronomers as 'Population I'. These rare but glittering youths outline most of the constellations, and have only been found in the spiral arms of galaxies where stars are actively produced.

THE INTERSTELLAR MEDIUM: DUST

Apart from the effect of perspective, our view of the galaxy is also affected by one of its minor constituents, dust. Tiny solid particles, typically 0.001mm or less across, make up only about one per cent of the mass of the galaxy, but they are concentrated in the galactic plane. Mixed in with the dust there is gas, mostly hydrogen, which comprises perhaps a quarter of the galaxy's mass. While gas is transparent, dust is not, and it is this dust that we see dividing the Milky Way in two along most of its length.

The division seems clearest, and the Milky Way broadest, in Sagittarius, which is low in the southern sky in the northern hemisphere summer (and in the northern sky in the southern hemisphere winter). The dust protects the interior of the clouds from the ultraviolet light from hot stars, but the clouds are largely transparent to infrared radiation, so they cool. As they do so, delicate organic molecules condense onto the cold grains, and molecules of hydrogen gas form. These molecules are detectable at radio wavelengths, and the dusty regions are known as 'molecular clouds'.

Most star formation occurs in dust clouds, which is why star-forming nebulae are there, and groups of young stars are scattered through the plane of the galaxy. When the gas-rich dust collapses in on itself, new stars are formed.

This high resolution image of the galactic centre was made at a wavelength of 6cm by the Very Large Array radio telescope in New Mexico. It covers a tiny area of sky two arc minutes on each side, one-fifteenth the apparent diameter of the Moon.

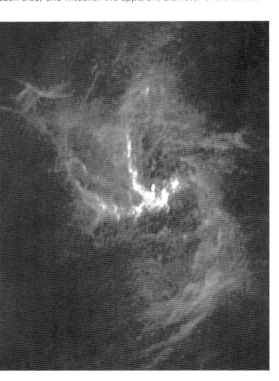

Above: The galactic centre is shrouded in dust, but it is surrounded by myriad stars forming the brightest part of the Milky Way.

Right: By contrast, the same camera, film and exposure records very few stars towards the south galactic pole.

The young stars then illuminate the clouds, producing spectacular nebulae, until the heat and light from the new stars drives the dust and gas away. As it disperses, a group of hot, young stars is revealed, prominent against a fainter background.

THE GALACTIC CENTRE

The bulge around the centre of the nucleus is the mass of old, cool and faint stars whose orbits carry them far above the plane. Known as Population II stars, they can be seen in vast numbers in Sagittarius. It is increasingly evident that the galactic centre is dominated by a massive black hole, a feature our galaxy has in common with many others. For now, the black hole is quiescent, but it may once have been more active, perhaps even a quasar in the early universe.

A DARK HALO

The most massive component of the galaxy can only be detected indirectly. Like other galaxies, Milky Way is surrounded by a substantial, dark halo, perhaps 10 times more massive than the 'visible' galaxy.

This 'dark matter' mass dominates the dynamics of galaxies. It cannot be seen at any wavelength but astronomers have deduced that it is there because of its effect on the speed at which they see other galaxies rotating. This 'missing mass' dominates the dynamics of clusters of galaxies as well as individuals. That it exists is not in doubt; only its nature and origins remain a mystery.

THE ZONE OF AVOIDANCE

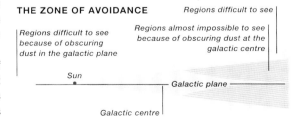

Regions difficult to see

Regions difficult to see because of obscuring dust in the galactic plane

Regions almost impossible to see because of obscuring dust at the galactic centre

Sun

Galactic plane

Galactic centre

The zone of avoidance was so named by Herschel, who noted an absence of faint extended objects (now known to be distant galaxies) in that direction.

TYPES OF GALAXIES

EDWIN HUBBLE CONFIRMED THAT THE MILKY WAY WAS ONE OF MANY GALAXIES. BUT NO-ONE IMAGINED THE GREAT DIVERSITY OF THESE COLLECTIONS OF STARS.

The Hubble classification of galaxies illustrated with real examples. Elliptical galaxies vary from E0 (circular) to E7 (elongated). Lenticular galaxies, S0, lie between elliptical and spiral galaxies. Spiral galaxies and barred spiral galaxies are classified into three types: (a) that have closely wound arms and large nuclei to (c) with loosely wound arms and small nuclei. Spirals of type (b) are in between the two. Most galaxies are irregular.

It was not until Edwin Hubble's discovery that M31 and M33 are, in fact, galaxies that scientists realized that there are different types of galaxies. This variety is reflected in their main distinguishing features; their mass, their luminosity and their shape.

THE HUBBLE CLASSIFICATION

The most widely used scheme of classification was devised by Hubble during the 1920s and revised by Sandage in 1961. It defines three major classes of galaxy: elliptical, spiral and irregular. There are also several sub-classes as illustrated in the famous 'tuning fork' diagram (below).

Howevever, not all of Hubble's conclusions turned out to be right. Hubble perhaps hoped that his diagram represented

THE LOCAL GROUP OF GALAXIES

The Milky Way has a retinue of faint spheroidal companions known as members of the Local Group. This group includes M31 – the Andromeda galaxy – and the Magellanic clouds. Some are so faint that they have only been discovered by careful electronic analysis of photographs. Among the brightest of the faint galaxies is the Leo I dwarf, discovered in 1950. Like most small groups the Local Group's brightest galaxies are those that look altogether more interesting than the bland ellipticals, the gas-rich spirals.

COMMON GALAXIES

Hubble wrongly believed that the commonest type of galaxies were the graceful spirals. Spirals are easy to see because they are usually bright, but the most numerous galaxies are faint dwarf elliptical galaxies, which are only detected on the deepest photographs.

Ellipticals seem particularly prevalent in the local universe. In rich clusters of galaxies, ellipticals are always the most luminous (and the most numerous) type.

One of the most massive ellipticals is M87 (NGC4486) in the nearby Virgo cluster, easily detectable at radio wavelengths. This galaxy is at least five times as luminous as the most luminous spirals and much

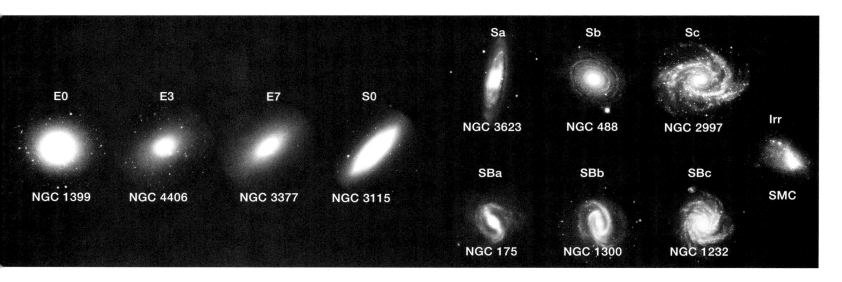

an evolutionary sequence because he used the terms 'early' and 'late' to describe ellipticals and spirals, though he added carefully 'without regard to their temporal implications'. It is now generally believed that there is no truth in this.

A UNIVERSE OF GALAXIES

The Milky Way is in the top league of massive spiral galaxies, with a total mass of about one thousand billion Suns. (Galaxies' masses are conventionally measured in terms of the Sun's mass.) The most massive spiral galaxy is the inconspicuous Malin-1, which is a spiral about 10 times more massive than the Milky Way. It has about the same mass as the biggest elliptical galaxies. The least massive galaxies are also ellipticals, and weigh in at a mere million solar masses or so.

bigger. Surrounding it on deep photographs are large numbers of much fainter 'dwarf' ellipticals, or dEs.

TYPES OF ELLIPTICAL GALAXIES

Elliptical galaxies range in shape from anywhere between a sphere to an American football. However, this classification is not straight forward because it often depends on the angle from which the galaxies are observed. An American football seen end-on would also look like a sphere. This means that some of the galaxies astronomers see as perfectly round (which are classified as E0) could also be elongated in shape when seen from a different perspective.

The visual classification of ellipticals extends through increasing elongated spheres, ranging from E1 to E6, with an almost cigar shaped E7. Galaxies with an otherwise smooth light distribution that are more elongated than E7

are known as S0s ('S-zeros') and these are also described as 'lenticular' (lens-like) from their edge-on appearance.

STELLAR MOTIONS

The shapes of the galaxies are governed by the dynamics of the stars of which they are made. In a perfectly symmetrical E0 galaxy the stars travel in randomly oriented elliptical orbits of various velocities around the nucleus, rather like an agitated swarm of bees around a hive. At a distance from the hive there are fewer bees. Rotation of the galaxy as a whole, if it occurs at all, is not significant.

The stars move in a similar way in an E7 galaxy, but their orbits are confined to narrow paths on a flattened plane, as though most of the bees prefer an orderly excursion along the same horizontal plane.

In S0s the stars form a true rotating disk, with the bees all moving in the same direction around the hive. However, within the disk there may be some stars closer to the nucleus that orbit out of the plane. These stars are visible as a central bulge when seen edge-on, and as a bright central flat disk while seen face-on. Distinguishing between them demands careful measurement of the distribution of

This elliptical galaxy, NGC205, is a one of the members of the Local Group of galaxies. It lies in the constellation of Andromeda, near the much more spectacular M31, the Andromeda galaxy itself. This image, taken from the Kitt Peak National Observatory, resolves individual stars in the galaxy.

Above: Active galaxy NGC5128 is identified as the source of intense radio and X-ray radiation known as Centaurus A. It is an elliptical galaxy cut across by a broad belt of gas and dust. Astronomers now know that NGC5128 is a 'cannibal' and is in the process of devouring a smaller galaxy.

Types of spiral galaxies 118–119 ►

Irregular galaxies 120–121 ►

The electromagnetic spectrum 140–141 ►

light, and untangling the dynamics of these seemingly simple systems is very difficult.

LIGHT AND DUST IN ELLIPTICALS

Ellipticals become gradually fainter from the bright nucleus to the dimmest outskirts. The light in these almost featureless galaxies is predominately that of red giant stars. The stars in ellipticals belong to Population II, which means that they are all very old. The gas and dust that is present is normally very tenuous.

Ellipticals with noticeable dust or young stars are generally classified as 'peculiar'. One of the nearest elliptical galaxies, NGC205 (see previous page), a companion of M31, falls into this class. It is close enough for astronomers to make out the individual stars and patches of dust. Another very peculiar elliptical with very prominent dust is the radio galaxy Centaurus A, NGC5128 (see previous page).

TYPES OF SPIRAL GALAXIES

Spiral galaxies are the second type of galaxy classified by Hubble. They fall into two main groups, again distinguished by their shape, and form the fork in the Hubble diagram.

Normal spirals are like the Andromeda galaxy and have arms which emerge from a central spheroidal bulge; they are subdivided in the Hubble scheme according to the spacing of the spiral arms.

Galaxies of type Sa have circular arms that are tightly wound; they generally have relatively little gas and dust and have large central bulges. In type Sc the arms are loosely wound, with plenty of gas, dust, star formation and a small central bulge. Type Sb have

M87 is one of the most massive known elliptical galaxies. It is the brightest galaxy in the Virgo cluster and is unusual in that it is surrounded by a very large number of faint globular clusters. It also has a bright jet that extends from the nucleus (not seen here).

This faint collection of stars is one of two similar dwarf elliptical galaxies known in Leo. The stars are all very old and there is very little interstellar dust and gas remaining.

M83 (NGC5236), in the southern constellation of Hydra, is one of the finest spiral galaxies in the sky.

intermediate characteristics, but the Hubble types are not rigid categories and many galaxies are in-between types such as Sab or Sbc.

In 'barred' spirals the curved arms seem to originate at the ends of an elongated bar, rather than a central bulge, and occasionally form a complete circle around it. Barred galaxies comprise about 20 per cent of all spirals and one of the finest examples is NGC1365 in Fornax (see over). These galaxies are likewise classified in the Hubble scheme by the openness or otherwise of their spiral arms and NGC1365 is type SBb The reason for the shape of the bar is not fully understood, but it probably reflects some instability that alters the orbits of the inner stars, forcing a more stable elongated central bulge.

Modern observation techniques have revealed a host of more subtle characteristics, and the Hubble scheme has been extensively modified to take them into account. M83, for example, has open spiral arms and looks like an Sc, but careful inspection of the nucleus shows it to have a small bar, so it is an SBc. The Milky Way is also suspected of being a barred galaxy.

STRUCTURE OF SPIRAL GALAXIES

The detailed anatomy of a spiral galaxy has already been examined in the Milky Way. Atlhough the band of the Milky Way is visible in our sky, little of its structure can be seen from the inside. As such, scientists have learned most of what they know about our galaxy from studying Milky Way-like spirals at a distance.

Most of them appear to have two more-or-less symmetrically curved arms that originate in a diffuse spheroidal bulge or an elongated 'bar'. The arms themselves are often clumpy, quite unlike the smooth envelopes of elliptical galaxies. In colour photographs the curved arms of spirals are bluish, while the regions around the nucleus are a pale yellow. This suggests that the light in the spiral arms comes from very hot, young stars that

belong to Population I, while that in the central regions comes from cooler, fainter, older stars of Population II. In fact, as you might expect, the situation is a bit more complicated.

Although most of the light of the spiral arms is produced by massive blue stars that are younger than the Sun, such stars are rare. Each of them, however, is extremely bright, up to a million times more luminous than the Sun, so relatively few are needed to make the spiral arms visible. These vigorous young stars have short lives, exploding as supernovae within a few tens of millions of years. They are greatly outnumbered by lightweight stars such as the Sun, which are born in much greater numbers, but whose light in the spiral arms is of little significance.

NGC891 is one of few galaxies seen edge-on, clearly revealing the dust lane confined to the galactic plane, and the central 'bulge' of old stars above and below the galactic disk.

Distribution of stars 120 ▶

Peculiar galaxies 121 ▶

The electromagnetic spectrum 140–141 ▶

NGC1365 is a spectacular barred spiral galaxy in the southern constellation of Fornax. Most of the other galaxies in this cluster are ellipticals.

PATCHES OF NEBULAE

Patches of nebulae, the source of young stars, are generally scattered in between the dark trails of dust and the glowing spiral arms that make these galaxies so visually attractive. This is not accidental, since young, short-lived stars are likely to be found near where they were born. The arms are clumpy because star formation is sporadic and bright clusters of stars are formed rather than individuals.

In nearby galaxies such as M83 (see previous page), groups of pink nebulae are dotted along the spiral arms. This is a particularly active star-forming galaxy and has hosted half a dozen supernovae in the last 70 years.

DISTRIBUTION OF STARS

Supernovae blast some of this star-forming material high out of the galactic plane but also compress the gas that remains, so a self-sustaining wave of star formation is driven through the galaxy, leaving a litter of stars in its wake. Since the hottest of them only live a short time, the pale blue arms soon fade away, leaving a much fainter but more numerous population of less massive stars in the inter-arm region. Thus the existence of spiral arms reflects the changing ratio of Population I to Population II with time, and as the glittering wave of star formation is driven around the galaxy it will encounter old stars that formed in previous cycles of birth and death.

DARK DUST LANES

Dust lanes occur between star-forming regions, which are tightly constrained by gravity and rotation, both within and towards the galactic plane in 'normal', undisturbed galaxies that do not have other galaxies interacting with them. This gravitational confinement contributes to the efficiency of star formation; dispersed dust and gas does not form stars. The narrowness of the dark lanes is clear in galaxies that scientists can view edge-on, such as NGC891 (see previous page).

IRREGULAR GALAXIES

Some galaxies do not fit neatly into either of the main classes and are described as 'irregular'. This is a mixed bag of unusual but often very interesting oddities and much about normal galaxies has been learned by studying what it is that makes irregular galaxies different.

The Magellanic clouds, the nearest galaxies to the Milky Way, are irregular. They are made up of a small and a large galaxy. Their oddness is the result of their interaction with our galaxy and with each other, which has stripped most of the hydrogen from the smaller galaxy and triggered intense star formation in the larger one. Radio observations trace the Magellanic stream of hydrogen in a vast arc around the Milky Way.

Irregular galaxies seem to be smaller than mainstream ellipticals and spirals but not all are interacting. The unusual galaxy NGC1313 (see below), for example, shows no sign of having a companion and is known as a 'starburst' galaxy

Above: The Large Magellanic cloud is the nearest galaxy to the Milky Way. Although classified as an irregular galaxy, scattered traces of a spiral structure can be seen.

Right: NGC1313 is a galaxy which has experienced recent and vigorous star formation. The clumps of young stars give this 'starburst' galaxy its chaotic appearance.

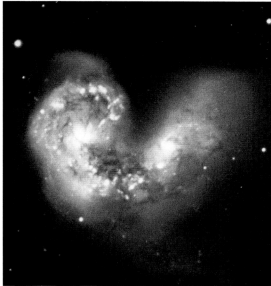

Many galaxies show signs of interaction with others, but few are quite as spectacular as the Antennae, a pair of colliding spirals, NGC4038–39. The black and white photograph (far left) shows huge curved sprays of stars that have been displaced by gravitational and tidal forces in the encounter, while the colour image (left) reveals brilliant clumps of young stars and pink star-forming regions hidden in the debris.

because a sudden rash of star formation gives it an irregular appearance. However, the underlying dynamics of the galaxy are not especially disturbed.

PECULIAR GALAXIES

It is a simple fact that, like people, most apparently normal galaxies have their oddities if you look hard enough. It may well be that truly normal galaxies are rarities, but often the abnormalities that make galaxies peculiar can be quite subtle. Although the range of peculiarities is almost endless the number of causes is limited to a few.

The most potent is interaction, or at its most devastating, complete absorption of one galaxy by another. The famous galaxy Centaurus A, NGC5128 (see page 117), detectable at radio wavelengths, is a fine example of galactic cannibalism: its dust lane is the remains of a gas-rich spiral that fell into it about a billion years ago.

Other peculiarities include: dust and young stars in elliptical galaxies; ellipticals with faint arcs of stars around them; galaxies with counter-rotating cores and disks, and spirals with anomalous arms. In almost all these cases the peculiarities are the result of interactions between two or more galaxies. The interacting pair of spirals known as the Antennae is one of the best examples, with gigantic faint arcs of stars seen on deep photographs and vigorous star formation (visible in the photograph above right).

THE HUBBLE DEEP FIELD

When the universe was much younger and the galaxies closer together, interactions were much more common. Although long-suspected from the anomalously high number of blue galaxies seen at great distance, we can now make this simple statement with some conviction from the solid evidence of spectacularly deep images from the Hubble Space Telescope, the Hubble Deep Field (HDF). In a tiny patch of sky a few arc minutes square we see many galaxies as they were when the universe was a fraction of its present age.

Almost all of the images we see in the HDF are galaxies, and they far outnumber local stars. Hubble and his

contemporaries would have been astounded. Most of these galaxies seem to be peculiar and many are undergoing star formation with a vigour that is not seen in nearby galaxies today, suggesting an excess of anomalously blue galaxies when the universe was a fraction of its present age.

By now many of the small galaxies seen in the HDF will have been absorbed by massive companions and the apparent normality of nearby galaxies today masks a turbulent past of which only traces remain.

Several lines of argument and years of observation from the ground and with orbiting telescopes strongly suggest that star formation increased gradually, reaching a peak when the universe was about one-fifth of its present

This part of the famous Hubble Deep Field image shows a part of the sky a few arc minutes across. Almost all of these fuzzy blobs are galaxies, most of them so distant that they are seen as they were billions of years ago.

Studying the early universe 128 ►

The Hubble Deep Field 128 ►

The electromagnetic spectrum 140–141 ►

Hubble's impact 144–145 ►

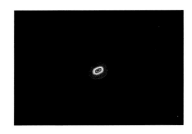

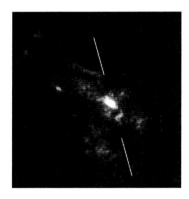

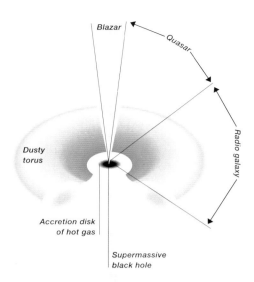

Blazar

Quasar

Radio galaxy

Dusty torus

Accretion disk of hot gas

Supermassive black hole

DIFFERENT VIEWS
It has long been believed that many of the properties of 'active' galaxies, blazars, radio galaxies and quasars could be explained by the existence of a torus of obscuring matter surrounding a black hole. This explains why the apparent nature of these objects varies with their orientation on the sky and from what angle we see them.

Top: A false-colour infrared image of blazar 3C279. Blazars are galaxies whose energy output increases 10 times or more over a period of a few months. It is thought that this may be the result of a star being sucked into the galaxy's central black hole.

Centre: Quasars are compact, extremely energetic light sources billions of light years away. Like blazars they are thought to be powered by black holes.

Bottom: Radio galaxies do not always have a visible counterpart, and can often be very different in appearance. For example, the radio map of Centaurus A is much more extensive than the visible galaxy.

age, between seven and nine billion years ago. This was also the time when some of the most extreme kinds of galaxies such as quasars are found.

QUASARS

Quasars are brilliant energetic nuclei of galaxies that are so distant they appear to be stars. One of the most remarkable surprises of this century was the discovery of radio waves from discrete sources in space. It was some time before the positional accuracy of radio observations was good enough to pinpoint the optical counterparts of these radio beacons, and when they were identified some of the most powerful looked just like stars in photographs.

Astronomers now realize that some of these sources, found at the very edge of the observable universe, are the most energetic objects known. Their discovery in the 1960s marked the beginning of a new era of high-energy astrophysics. Astronomers soon discovered that the most extreme of these objects emitted 10 million times more radio energy than the Milky Way and a hundred times more light. Many radiated energy at X-ray, ultraviolet and infrared wavelengths as well.

Once their distance was established, by means of their redshifted emission lines, it became apparent that though these were very energetic objects, they were remarkably small. This implied the source was only a few light days or light weeks wide, itself an amazing revelation.

The only source that might produce this kind energy with such a broad and distinctive spectrum of radiation in so small a volume is matter pouring into a black hole. But when quasars were first discovered, black holes were only a theoretical notion, so bizarre as to be unlikely.

Fortunately for the theorists, evidence for the existence of black holes was growing. Other phenomena were discovered whose nature could only be explained by black holes: the powerhouses of supernova remnants, exotic binary stars and objects with active galactic nuclei (AGN). AGNs, found in nearby galaxies, are similar to quasars in that they give off a large amount of energy, but they are less extreme.

DIFFERENT VIEWS

Around the nucleus of quasars and AGNs is thought to be a thick torus – a doughnut-shaped ring of gas and dust. It is from this torus that material spirals into the central black hole from 'an accretion disk' emitting vast amounts of energy as it does so.

At first there seemed to be several different kinds of quasars and AGNs. Now it seems that much of the perceived difference is the result of the way we see these galaxies in the sky.

Seen edge-on, the accretion disk is hidden at optical wavelengths and astronomers can only see jets of material ejected at right angles to the obscuring torus. With a more end-on view the radiation from the accretion disk dominates everything and the quasar or AGN looks quite different. When observing at angles in-between, astronomers see the radiation from the accretion disk bombarding the inner wall of the torus and producing a distinctive spectrum of absorption and emission lines. At infrared wavelengths astronomers might glimpse the accretion disk itself.

Of course there are other differences between individual objects and not all tori have the same dimensions, but this broad picture seems to be true of most quasars.

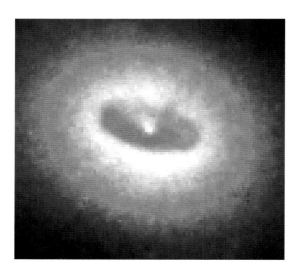

In 1992, the Hubble Space Telescope discovered a spiral-shaped disk of dust that fuels a massive black hole in the centre of the galaxy NGC4261. Astronomers calculate that this object is not much larger than our solar system, but has 1.2 billion times the mass of our Sun.

The Hubble Space Telescope confirms that quasars are galaxies, often members of a cluster, and has produced the first clear picture of a toroidal disk around an AGN.

GALAXY CLUSTERS

In general, galaxies are gregarious and isolated examples are rare. Whatever the mechanism is that creates galaxies it seems to produce them in clusters rather than individuals, rather like the formation of stars.

We are in a cluster of about 30 galaxies, the so-called 'Local Group', where the Milky Way and the Andromeda galaxy, M31, are the major galaxies and the rest are small fry. The Local Group is typical of small clusters in that it is dominated by gas-rich spirals.

Clusters containing a greater number of galaxies have increasing numbers of ellipticals: most members of the richest clusters in the local universe are ellipticals. The nearest medium-rich cluster is 50 million light years away in Virgo; many of its members are bright enough to have been entered into Messier's catalogue 200 years ago.

THE STRUCTURE OF THE UNIVERSE

The galaxies in the spectacular HDF image seem to be scattered at random across the picture. The photograph is a deep but very narrow probe through the structure of the universe. Astronomers can tell from their size that most of the HDF galaxies are at a great distance. But how distant? Even in wide-angle pictures they are unable to see them in three dimensions.

They can, however, 'see' the third dimension, distance, by measuring redshift. The expansion of the universe draws the galaxies apart, and astronomers perceive the expansion as though everything is moving away from us. The more distant the galaxy the higher the velocity of recession.

If the universe is expanding uniformly, as we believe, redshift is related to distance by Hubble's constant, which predicts distance from the rate of recession of galaxies. The value of this much-discussed parameter is now widely accepted to be about 75km/sec/Mpc, which means that for every 3.26 million light years of distance the speed at which the galaxies are moving apart from each other increases by 75km per second. This seems true in all directions, except for small local variations. The most distant galaxies are receding at hundreds of thousands of kilometres per second, almost at the speed of light itself.

When astronomers measure the redshifts of galaxies over a wide-angle slice of sky we find that they are not scattered at random distances. Instead they form vast three-dimensional sheets and clumps, as though the material from which they were made was a tenuous cosmic foam, with 'galaxy stuff' concentrated in the walls and intersections of the bubbles, with great voids in between. This explains the distribution of the raw materials from which galaxies were formed but it poses a puzzle. Why is the universe not uniform?

SUPERCLUSTERS

The Local Group is part of the local 'supercluster', a more-or-less disk-shaped aggregation of lesser galaxy

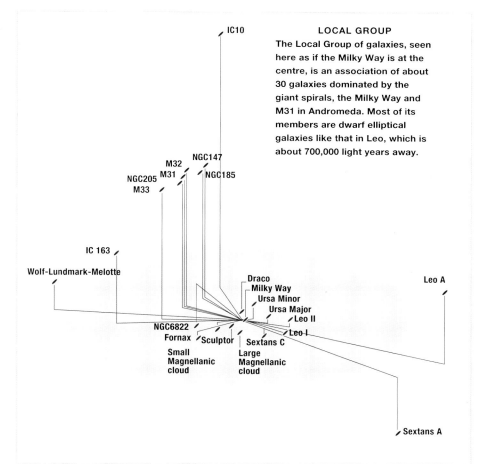

LOCAL GROUP
The Local Group of galaxies, seen here as if the Milky Way is at the centre, is an association of about 30 galaxies dominated by the giant spirals, the Milky Way and M31 in Andromeda. Most of its members are dwarf elliptical galaxies like that in Leo, which is about 700,000 light years away.

clusters over 60 million light years wide. Superclusters are joined together by filaments and sheets of clusters of galaxies. The voids in between seem to be almost completely empty.

One of the relatively close concentrations of galaxies is known as the 'Great Wall' which shows up as a 500 million light year long strip about 15 million light years thick running across the northern sky.

These observations were made by measuring the redshifts of thousands of galaxies one at a time. Now instruments such as the Anglo-Australian telescope's 2° field (in the southern hemisphere) and the Sloan Digital Sky Survey (in the northern hemisphere) will pursue a much more vigorous programme to measure the distances of millions of ever more distant galaxies. The next great cosmological adventure is about to begin.

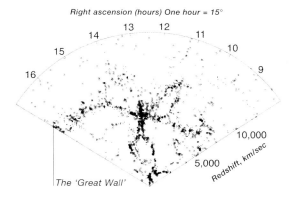

Right ascension (hours) One hour = 15°

The 'Great Wall'

Astronomers cannot determine an accurate distance for a galaxy by its appearance, but because the universe is expanding, the more distant a galaxy is, the greater is its velocity of recession. This 'redshift' can be measured by the displacement of lines in the spectra of the galaxies. This is how the third dimension is found in a thin slice of the sky illustrated here. A particularly striking feature is the 'Great Wall', a sheet of galaxies 500 million light years long.

COSMOLOGY

Cosmology is the study of the universe as a whole. The universe is everything: stars, galaxies, planets. Cosmologists attempt to explain why we see the universe as we do today, how it came into being and how it will finally die. Many astronomical observations help unravel the secrets of the universe. It is like a huge detective story where no-one knows who perpetrated the crime.

The most commonly accepted theory of how the universe began is the Big Bang, where everything, matter, energy, radiation and the forces of nature, all came into being at the same time. Time also began with the Big Bang, so it is meaningless to ask what happened before the universe was born. The Big Bang theory explains many aspects of the universe today, but there are still many unanswered questions.

1 ▶ *Time: 0. The universe begins: space and time are created. All matter and energy comes into being.*

8 ▶ *Time: approximately 15 billion years (today). Temperature: 3K. Humans evolve and study the universe.*

THE HISTORY OF THE UNIVERSE
Everything that exists in the universe was created in the Big Bang 15 billion years ago. All the different kinds of subatomic particles were formed in the first minute; the simplest atoms after 300,000 years. Only after one billion years did matter begin to clump together to start forming galaxies.

7 ▶ *Time: one billion years. (10^9 years). Temperature: 20K. The matter in the universe begins to clump together to form galaxies in which stars are formed.*

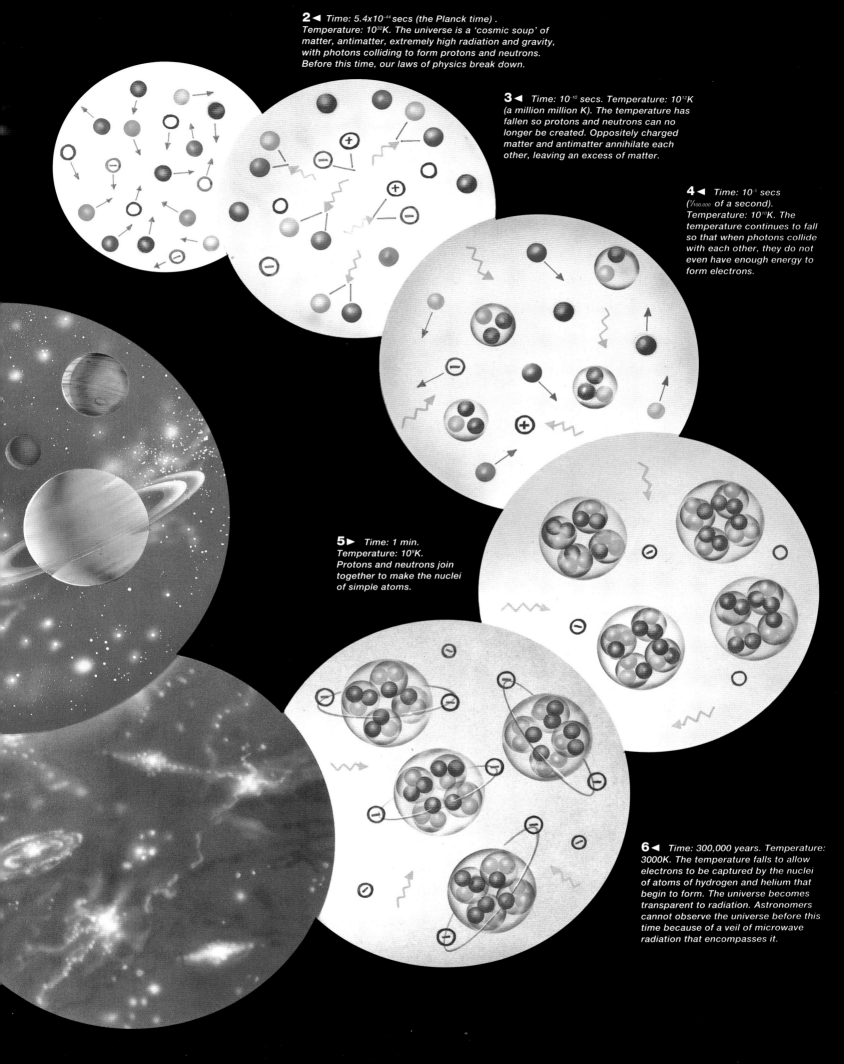

2 ◀ *Time: 5.4x10^{-44} secs (the Planck time).*
Temperature: 10^{32}K. The universe is a 'cosmic soup' of
matter, antimatter, extremely high radiation and gravity,
with photons colliding to form protons and neutrons.
Before this time, our laws of physics break down.

3 ◀ *Time: 10^{-10} secs. Temperature: 10^{12}K*
(a million million K). The temperature has
fallen so protons and neutrons can no
longer be created. Oppositely charged
matter and antimatter annihilate each
other, leaving an excess of matter.

4 ◀ *Time: 10^{-5} secs*
(1/100,000 of a second).
Temperature: 10^{10}K. The
temperature continues to fall
so that when photons collide
with each other, they do not
even have enough energy to
form electrons.

5 ▶ *Time: 1 min.*
Temperature: 10^{9}K.
Protons and neutrons join
together to make the nuclei
of simple atoms.

6 ◀ *Time: 300,000 years. Temperature:*
3000K. The temperature falls to allow
electrons to be captured by the nuclei
of atoms of hydrogen and helium that
begin to form. The universe becomes
transparent to radiation. Astronomers
cannot observe the universe before this
time because of a veil of microwave
radiation that encompasses it.

AFTER THE BIG BANG

OUR UNIVERSE, WHICH IS MADE UP OF MANY
GALAXIES, HAS BEEN EXPANDING EVER SINCE THE
BIG BANG. BUT WHAT WILL BE ITS ULTIMATE FATE?

Although astronomers are stilll a long way from fully understanding the universe, they do have an idea of how it came to be formed, its broad structure and the nature of its dynamics. From this they have been able to put forward a number of theories regarding its ultimate fate.

THE EVER-GROWING UNIVERSE

Vesto Slipher was one of the first to use spectroscopy to determine the physical and chemical state of a galaxy's constituents, early in the 20th century. At the time, astronomers were unsure about the nature of galaxies; some thought they were clouds in our own galaxy and others (correctly) thought they were distant collections of stars.

What Slipher showed was that a galaxy's spectral lines were displaced from where they were expected to occur. Aware that the Doppler shift could change the wavelength of spectral lines (thereby displacing them) Slipher concluded that the galaxies were moving very fast, most of them away from the Milky Way. Thus, the concept that a galaxy possesses a recessional velocity was born.

REDSHIFT

The astronomer who continued this line of work was Edwin Hubble. He measured the distances to some of these galaxies by comparing a type of nearby pulsating star, known as a Cepheid variable, to others of that kind that he found in those distant galaxies.

When Hubble plotted their distances he noticed that, with a few exceptions, the farther away a galaxy was, the greater the displacement of the spectral lines and hence the greater the recessional velocity.

The shift is caused by the wavelength of the spectral line being lengthened as the galaxy moves away, and hence the phenomenon became known as the redshift, because red light has a longer wavelength than blue light. So the farther away a galaxy is, the faster it is moving away

from the Earth, the greater its redshift, and astronomers came to realize that the entire universe is expanding.

Hubble went on to calculate how fast the universe was currently expanding. In order to visualize the expansion of the universe, imagine a model of the universe with galaxies joined together by matchsticks. Then imagine the matchsticks growing longer. In this way the model of the universe expands, carrying the galaxies farther and farther away from each other.

In crowded regions of the universe, the gravitational force between galaxies overcomes the expansion of space. This accounts for the handful of galaxies that do not show a redshift from the Earth because they are each gravitationally bound to our galaxy, the Milky Way.

THE GENERAL THEORY OF RELATIVITY

At around the same time as Hubble was working, physicists were getting to grips with Einstein's mathematical description of gravity: the General Theory of Relativity. In General Relativity, space is rather like a framework within which the stars and the galaxies are suspended. Time is also a part of this framework, which is known as the space-time continuum.

Einstein encountered considerable difficulties with his equations because they showed this framework should be expanding or contracting, and not be stationary. It was widely believed at the time that space was static, that objects could move through space, but space itself was just a 'nothingness'. To comply with this, Einstein introduced into his equations a mathematical term, known as the cosmological constant, to counteract the pull of gravity and hold the universe static.

When Hubble published his redshift results, the problem was solved and Einstein's original equations, without the additional constant, were proved correct.

BACK IN TIME

With the discovery of the expanding universe, scientists realized that in the past, the constituents of the universe must have been very much closer together. Georges Lemaître was the first to suggest that the universe began in a Big Bang (although he called it a primeval atom).

Theoreticians have since spent a lot of time wondering about what conditions must have been like during those early times. One of their most interesting conclusions is that the four forces of nature would all have acted in the same way.

THE FUNDAMENTAL FORCES OF NATURE

For most of its life, four fundamental forces of nature have shaped the universe. Gravity governs the way massive objects attract each other. Electromagnetism determines the way charged objects interact with one another whilst the strong and the weak nuclear forces determine the structure of atomic nuclei and the interactions they can have with other non-charged particles.

In 1983 the European Centre for Nuclear Research (CERN) mimicked the conditions of the universe when it was only 10^{-11} seconds old (one hundred thousand millionths of a second) and confirmed a theory that the

Edwin Hubble discovered a Cepheid variable in the Andromeda galaxy (M31) in 1923. From this observation, astronomers were able to work out that what had previously been thought to be the Andromeda nebula was, in fact, a huge system of stars lying a long way outside the Milky Way.

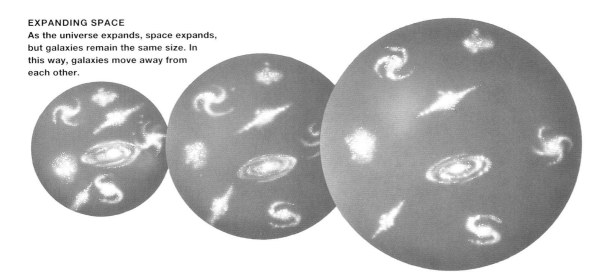

EXPANDING SPACE
As the universe expands, space expands, but galaxies remain the same size. In this way, galaxies move away from each other.

A true-colour image of faint blue galaxies at the edge of the observable universe, photographed by the William Herschel Telescope on La Palma in the Canary Islands. The bulk of the faint blue galaxies are probably in their first phase of star formation and the sheer number of them means that they must have been much closer together than is the case now.

Studying the early universe 128 ▶

The electromagnetic spectrum 140–141 ▶

Spectroscopy 142 ▶

Many different techniques and instruments are used to further astronomers' understanding of the universe. To study how matter might have behaved in the early universe, physicists try to recreate those conditions in particle accelerators (massive insutruments in which atoms are smashed into each other at incredibly high speeds), so that they can examine how the particles interact.

electromagnetic and the weak nuclear forces would have acted as one.Scientists now believe that the strong nuclear forces would have unified under the conditions that existed at 10^{-35} seconds after the Big Bang.

These Grand Unified Theories are as yet unproved but some physicists have gone even further and believe that gravity, too, would have been indistinguishable at the universal age of 10^{-43} seconds. This time is known as the Planck time.

STUDYING THE EARLY UNIVERSE

The study of our cosmic origins has been made possible by the fact that light, as well as the other forms of electromagnetic radiation, travels with a finite speed. On Earth, distances are so short that light appears to travel instantaneously. In space, however, distances can be so enormous that the light from an event can take many – sometimes millions or billions of years – to reach the Earth.

By using large telescopes to collect the light that has been travelling for these amazingly long periods of time, astronomers can see images of the way the universe looked, many millions and billions of years ago.

THE HUBBLE DEEP FIELD

One of the most spectacular images in modern astronomy is known as the Hubble Deep Field. This was taken in December 1995 when the Hubble Space Telescope observed the same small area of space for 10 consecutive days. It found over 3000 extremely faint galaxies in the very distant reaches of the universe; all at different distances and different stages in their evolution. Some appear as they did when the universe was only a few billion years old.

Unfortunately, it is not possible to look even farther into space and see the Big Bang. A 'barrier' exists that effectively blocks our view of events earlier than 300,000 years into the history of the universe. The barrier is an all-encompassing swathe of microwave radiation. It has been produced because, before an age of 300,000 years, no atoms could exist in the universe.

A HOT BEGINNING

The temperature of space is governed by the energy carried on photons of electromagnetic radiation. These were so hot for 300,000 years that they prevented atoms from forming, by colliding with the nuclei and especially the electrons. With lots of electrons being scattered all over the universe, collisions between them and the radiation were a frequent occurrence.

As the universe expands, the electromagnetic radiation travelling through space is stretched, causing it to redshift to longer wavelength radiation. Thus the energy carried by each photon decreases.

Arno Penzias (left) and Robert Wilson were working for the Bell Telephone Company when they discovered the cosmic background radiation with a radio telescope.

Around 300,000 years after the Big Bang, the energy of the photons had dropped so much that they could no longer prevent electrons binding with atomic nuclei. The number of collisions also dropped because as the electrons were gathered up by the nuclei they ceased to be obstructions to the propagation of the photons.

At this point the universe changed dramatically. Whereas, previously, the radiation had travelled only a short distance before interacting with matter, now it could easily travel the entire width of the universe with only a small possibility of an interaction. Astronomers refer to this watershed as the decoupling of matter and energy. It happened when the temperature of space was about 3000K.

Since that time the universe has expanded by a factor of 1000 and so the radiation released at the decoupling as gamma rays has been stretched out and can now be seen as microwaves, which heat space to a temperature of just 3K.

MICROWAVE BACKGROUND RADIATION

The cosmic microwave background radiation is one of the important cornerstones on which the Big Bang theory rests. Although Lemaître had been the first to conceive the idea of a beginning to the universe, it was George Gamow, in the 1930s, who was the first to think seriously about the consequences.

Gamow was seeking to explain the origin of the chemical elements. In the process, he realized that if you compressed the contents of the universe they would heat up, in the same way as air compressed in a bicycle pump. The heat would allow an initial mixture of protons, neutrons and electrons to collide and stick together, forming atoms.

He collaborated with Ralph Alpher and Robert Herman, who realized that the Big Bang should have filled the universe with 'left-over' radiation, calculating its temperature to be 5K. Collaborating on theories about chemical elements, however, the team were not entirely successful and, subsequently, the work on the background radiation languished.

In the 1960s Bob Dicke realized that it should be possible to detect and measure the radiation, if it existed. His collaborators David Wilkinson and Peter Roll set about constructing a telescope to do the job whilst James Peebles began theorizing about what such a discovery would tell us about the early universe.

As the team neared completion of their telescope, in 1964, two other radio astronomers working for the Bell Telephone Company, Arno Penzias and Robert Wilson, detected a persistent signal with a radio telescope they were using. At first they thought it was a problem with the instrument but, when they consulted with Dicke and his group, they realized it was the 'left-over' radiation.

The discovery opened up the study of the early universe and Penzias and Wilson were awarded a Nobel Prize in 1978.

MAPPING THE UNIVERSE

In 1989 NASA launched the Cosmic Microwave Background Explorer (COBE) to study this radiation. By measuring the temperature of the radiation, COBE produced a map of the universe showing how it looked just 300,000 years after the Big Bang. Instead of stars and galaxies, the universe was simply filled with large clouds of gas that were denser in some places than others. These gas clouds were just starting their collapse into objects such as stars, galaxies and clusters of galaxies.

In the intervening few billion years, these clouds became the young galaxies such as those in the Hubble Deep Field. These, in turn, have evolved into the types of galaxies that are visible in the night sky through small telescopes.

THE FATE OF THE UNIVERSE

The universe cannot continue to expand at the same rate forever, because all the galaxies attract each other through their gravity. This has the effect of slowing down the

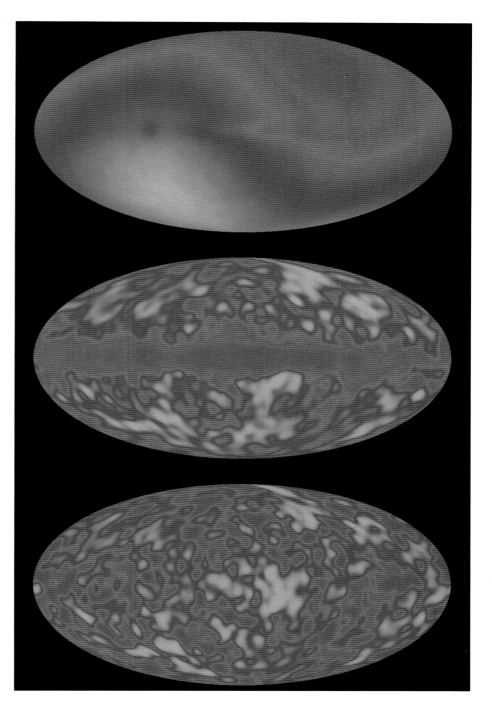

These images show 'all sky' maps taken in the microwave part of the electromagnetic spectrum by the COBE satellite. The pink regions show hotter areas than the blue. The top image show the Earth's own movement through space. The centre image shows the movement of our Local Group of galaxies and the bottom image shows the uniformity in the actual cosmic background radiation.

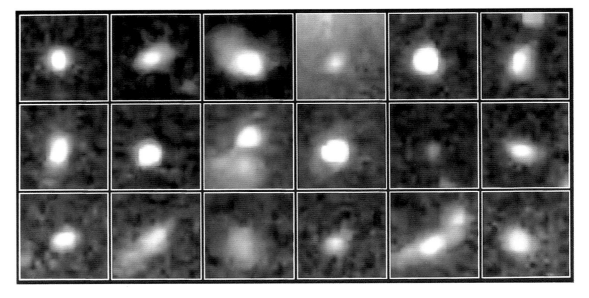

A sample of galaxies and quasars from the Hubble Deep Field image. Each object is 2000–3000 light years across. They are so distant that their radiation has been redshifted to be seen as visible light.

expansion. Over the entire history of the universe, the rate of its expansion (the Hubble constant) will steadily change its value: being higher in the past and lower in the future.

The rate at which the expansion decreases is a long sought after quantity in cosmology and is known as the deceleration parameter. It relies solely on the size of the universe and the density of matter in it. Once the deceleration parameter is known, the future changes of the Hubble constant can be calculated accurately and astronomers can answer the question: will the expansion ever stop?

The density of matter in the universe is the critical quantity that will decide this question. Too little matter and the universe will expand forever, too much and it will eventually stop and then the universe will collapse.

THE OPEN UNIVERSE
The case in which the universe expands forever is known as the open universe. In this instance, the galaxy clusters continue to get farther and farther away from each other. Galaxies themselves exhaust their supplies of star-forming dust and gas to become huge dark objects containing nothing but stellar remnants such as neutron stars and white dwarfs.

Eventually these ghostly galaxies, within their own gravitationally bound clusters, will collide and coalesce to become enormous black holes.

THE CLOSED UNIVERSE
If the universe is destined to collapse, however, circumstances are very different. The closed universe, as this case is called, reaches a maximum size and then begins to shrink. Instead of the expansion causing a redshift, the collapse will blueshift radiation travelling through the universe and everything will become hotter.

Eventually, galaxies will merge and the whole universe will simply become a mass of stars. Some stars will collide but many more will evaporate into space. This bizarre fate will be caused by the blueshifting of the cosmic microwave background radiation to a temperature in excess of that found on the surface of a star. As the temperature rises still further, the decoupling of matter and energy will be reversed as electrons are stripped from atoms.

Eventually, the universe will exist in a state very similar to that found in the Big Bang, before disappearing back into the nothingness from which it is assumed it came.

THE FLAT UNIVERSE
Helping astronomers to determine the eventual fate of the universe is a key prediction of the theory of inflation, which states that the universe suddenly expanded at an enormous rate. Inflation was caused by the break-up of the grand unified force into the strong nuclear force and the weak nuclear force at 10^{-35} seconds after the Big Bang.

The exponential expansion of space that inflation caused should have balanced the universe precisely between the opposite possibilities of open or closed. This very special case is known as a 'flat' universe and states that the universe contains just enough mass to stop it expanding after an infinite length of time has passed.

FATE OF THE UNIVERSE
Cosmologists think that the universe is following one of three fates: closed, flat or open. A closed universe will eventually stop expanding and collapse back in on itself, a flat universe will eventually stop expanding; and an open universe will expand for ever.

Distance between galaxies

The open universe

The flat universe

The closed universe

Time

The big crunch

The Big Bang

For all practical purposes, however, the fate of a flat universe is the same as that for an open one. The density of a flat universe is known as the critical density.

At present, astronomers' best observations suggest that only about 1 to 10 per cent of this critical density ismade up of ordinary atoms, known as baryonic matter. The other 90–99 per cent is known as the missing mass problem and is thought to be non-baryonic matter, or 'dark' matter.

THE MISSING MASS

Astronomers and particle physicists are searching very hard to try to discover the exact nature of this non-baryonic matter. In many instances it can be inferred by the gravitational effect it has on the way galaxies rotate and orbit each other in clusters.

Unfortunately, it still evades direct detection with even the most sophisticated equipment, despite the fact that a number of potential particles were predicted by the Grand Unified Theories.

THE SHAPE OF SPACE

According to Einstein's Theory of General Relativity, the presence of matter causes space to become curved. It is similar to putting a heavy ball on a rubber table top.

The overall curvature of the universe is therefore dictated by the density of matter it contains and this shape is correlated to each of the three eventual fates (see below). If the universe were two dimensional instead of three, the space-time continuum of an open universe would possess negative curvature and look like a saddle. A flat universe would be exactly flat whilst a closed universe would possess positive curvature and a two-dimensional slice would look like a hemisphere.

Each of these curvatures should affect how galaxies in the far distant universe appear to us. Some astronomers are now counting the number of far-off galaxies, hoping the results, will allow them to deduce the overall curvature of space.

Above: These faint blue patches of light, 11 billion light years away, are thought to be the building blocks for today's galaxies. Containing dust, gas and a few billion stars, these galaxies are close enough to merge and become larger.

THE SHAPE OF SPACE

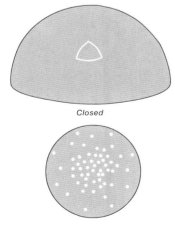
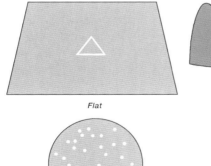
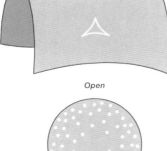

Closed

Flat

Open

The fate and the curvature of the universe are determined by the density of matter contained within it. If astronomers can ascertain the universe's shape, they can also determine its future. In order to do this, some astronomers are plotting the apparent distribution of distant galaxies. For each different shape (and so fate), the galaxies appear in a different pattern when plotted onto a flat sheet.

The great age debate 132–133 ▶

The electromagnetic spectrum 140–141 ▶

THE GREAT AGE DEBATE

In the 1920s, Edwin Hubble discovered that most galaxies are moving away from us. The farther away they are, the faster they are moving, indicating that the universe is expanding. The rate of this expansion is known as Hubble's constant, but an accurate figure for this ratio still eludes astronomers. A large value for the constant would imply a rapid expansion and a young universe, while a small figure would imply a slow expansion and an older universe.

BY DR PATRICK WALTERS

Edwin Powell Hubble was the American astronomer who discovered in the 1920s that the universe was expanding.

The Andromeda galaxy is a large spiral over two million light years away. Edwin Hubble's observations of this galaxy showed for the first time that it was a huge system of stars lying outside our own galaxy.

❝ A new era in cosmology began when Edwin Hubble (1889–1953) turned the new 100-inch telescope at Mount Wilson on the heavens. By photographing the outer spirals of the Andromeda nebula in 1923, he revealed them to be dense swarms of stars arranged in a pattern similar to that of our own Milky Way. Among these stars Hubble identified several Cepheid variables, which he studied in order to find their periods of variation in brightness. By doing so, he was also able to determine their luminosity, which in turn allowed him to work out how far away they were. He was able to show that Andromeda lay at an immense distance of almost one million light years. It is now known that the Andromeda galaxy is at a distance of more than two million light years.

Beyond it lay other, more distant galaxies with elliptical, irregular or spiral patterns. By noting that the more distant galaxies possessed greater redshifts, and thus were moving away faster, Hubble was able to make his most revolutionary announcement ever: the universe was not static, but expanding. This rule is known as Hubble's law and Hubble's observations eventually led to the formulation of the Big Bang theory.

Hubble's discoveries had many ramifications for cosmologists. If the universe is expanding, it must have been much smaller in the past, implying that it had a beginning; if you could calculate the rate of expansion, you could eventually calculate the age of the universe itself, based upon that rate. This rate of expansion is currently represented by an unknown figure known as Hubble's constant; larger values for this figure imply a rapid expansion and a younger universe, while smaller figures imply a slow expansion and an older universe. Not surprisingly, the value of this constant has been the subject of controversy and continual revision. The problem arises from the long chain of steps involved in deriving an exact

Instead of providing hard and fast answers the new technology revealed that the stars were older than the universe!

distance scale of the universe: if a galaxy moving away at a specific rate is found to be farther away than previously thought, it must have taken more time to reach this distance, implying a greater age for the universe.

The Belgian astronomer Georges Lemaître suggested that an explanation of Hubble's law might lie in an explosive creation of the cosmos. Early calculations of Hubble's constant suggested that the time elapsed since the creation event was surprisingly short. By the late 1920s, geologists had made the first estimates of the age of the Earth using radioactive materials embedded in rocks – and their findings implied that the Earth was older than the universe!

Observations with the new 200-inch telescope on Mount Palomar in California radically revised the estimated value for Hubble's constant. In 1952, Walter Baade realized that there were two types of Cepheid variables, and his revised measurements effectively doubled the estimated size and age of the universe. In 1958, Allan Sandage showed that another of Hubble's assumptions used in determining the distance to objects was also incorrect. These findings increased the estimate of the age of the universe well beyond the age of the Earth. By the late 1960s, Sandage and Gustav Tammann had published a value for Hubble's constant that suggested that the universe could be as much as 20 billion years old.

Still the controversy raged. At an astronomy conference in 1968, Gerard de Vaucouleurs issued a dramatic challenge to Sandage's work. He pointed out that the Milky Way galaxy was part of a huge 'supercluster' centred on the Virgo cluster of galaxies, and that cosmic expansion was distorted in this region due to the gravitational attraction of these galaxies to one another. He invented a series of distance 'stepping stones' and concluded that the universe was only half as old as Sandage had indicated.

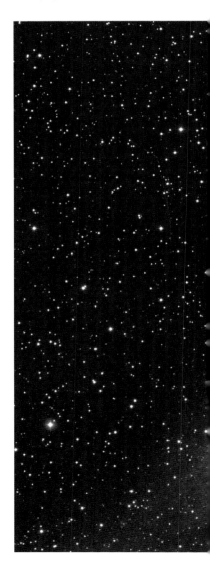

A new generation of astronomers using new electronic detectors, telescopes and techniques began to obtain results that supported the ideas of de Vaucouleurs, but instead of providing hard and fast answers, the new technology revealed that the stars were older than the universe! The oldest stars in the Milky Way galaxy occur in huge groups called globular clusters. An analysis of the relative evolution of different stars in these clusters indicates an age of about 15 billion years.

The Hubble Space Telescope was used to locate and isolate Cepheid variable stars in the Virgo cluster of galaxies (which is much farther away than any member of our local group), and this allowed scientists to determine reliable galactic distance measurement. In 1994, Wendy Freedman and her colleagues published the first results. The galaxy M100 in the Virgo cluster indicated a high value for Hubble's constant – which meant that the universe was less than 12 billion years old. Further observations of galaxies in the Virgo and Fornax clusters, as well as the neighbouring galaxies M101 and M81, confirmed these findings, but the age enigma remained unresolved.

In 1997, astronomers had their first opportunity to analyse data obtained by the European Space Agency's Hipparcos satellite. Hipparcos had measured the distances to stars with unprecedented accuracy. The data included observations for the first direct measurements of Cepheid variables. The result was unexpected.

Cepheid variables – the 'standard candles' of cosmology – were brighter and farther away than had previously been thought. This meant that the estimated value of Hubble's constant had to be decreased by about 10 per cent, and this lead to a corresponding increase in the assessment of the age of the universe.

The revised distance scale meant that the Milky Way's globular clusters, used to determine the ages of stars, were also farther away and brighter than previously estimated. This led Michael Feast and his team to realize that the oldest stars were only about 11 billion years old – younger than the revised estimate of the universe's age. Only time will tell if this revision has finally settled the century-long quest to determine Hubble's constant and the age of the cosmos. "

HUBBLE'S CONSTANT
Hubble discovered a direct relationship between how far distant objects are away from us and how fast they are receding: the farther away something is, the faster it is moving. If an object one megaparsec away is receding at 50km/sec, something two megaparsecs away will be receding at 100km/sec. In this case, Hubble's constant would be 50km/sec/megaparsec.

Once astronomers have found out how quickly objects are receding and how far away they are, they can then use their value for Hubble's constant to calculate how long they have taken to get there, and so work out how old they are, which leads to an estimate of a minimum age for the universe, which has to be older than the oldest objects contained within it.

PROFESSIONAL ASTRONOMY

Ever since humans walked on the Earth they have looked up and wondered about the heavens. Very gradually the secrets of the universe are being revealed as astronomers examine the sky with larger and more powerful instruments.

It has only been in the last 50 years that some windows of the universe have opened. With today's technology, and the ability to send spacecraft above the atmosphere, professional astronomers now examine the heavens in gamma rays, X-rays, ultraviolet and infrared radiation, and radio waves – right across the electromagnetic spectrum.

A host of intriguing objects has been discovered by viewing at these wavelengths; some questions have been answered and new ones have had to be asked.

WINDOWS TO THE SKY

THERE ARE MANY WAYS OF LOOKING AT THE UNIVERSE AND WITH EACH NEW ASTRONOMICAL DISCOVERY, A LITTLE MORE OF IT COMES INTO OUR GRASP.

GEORGE ELLORY HALE
George Ellory Hale (1868–1938) was not only an extremely successful astronomer but a man who oversaw the building of many telescopes, including three solar telescopes on Mount Wilson and the 5m telescope on Mount Palomar in the United States. Hale was also responsible for developing the spectrohelioscope, which monitored the effects in the Sun's chromosphere.

The Anglo-Australian Telescope at Siding Spring in New South Wales, Australia. The 4m-plus mirror was at the leading edge of technology in the 1970s and this has been an invaluable research instrument in the clear southern skies.

Professional astronomy has come a long way since Galileo first turned his telescope to the sky. However, there are various obstacles that astronomers strive to overcome by finding new windows out to the universe.

FARTHER AND FAINTER

Larger telescopes allow you to do two things. Firstly, they allow you to see fainter objects. Since stars and galaxies appear fainter the farther away they are, large telescopes enable you to see 'deeper' into space. Secondly, a larger telescope offers better resolution – the ability to see fine detail. This allows you to learn more about the object you're looking at. It is just like seeing a sign that is too far away for you to make out the letters; with a telescope you can read the message.

THE QUEST FOR APERTURE

The quest to view the universe has driven fantastic improvements in telescope size and design. The size of telescopes available to astronomers is limited by the engineering capabilities at any given time. By the turn of the 20th century, telescopes with apertures of a metre or so (today considered to be modest-sized instruments) were being constructed around the globe.

A telescope's aperture is the diameter of the lens or mirror that is pointed at the sky. By 1918, a 2.5m telescope had been completed on Mount Wilson, followed in 1947 by a telescope twice that size on Mount Palomar: the famous 5m Hale telescope.

During the 1960s and 1970s, several 4m telescopes were built, two of the most famous being those at Kitt Peak in the United States and Siding Spring in Australia. These instruments have since been dwarfed by the next generation of 8–10m class telescopes but are likely to remain invaluable as research tools.

While the largest telescopes are the most penetrating, the bulk of the routine, follow-up astronomical research (and discoveries) is carried out using smaller instruments.

THE WIDER VIEW

Increasing resolution and light-gathering power tends to have an unfortunate side-effect: reducing the field of view. Paradoxically, the more you see, the less you see. This phenomenon is well known to anyone who has ever looked through a pair of binoculars or a telescope: if you point binoculars at a tree, you might see individual leaves, but you will no longer see the whole tree.

The astronomer is like a tourist: there's so much to see and so little time to see it, so a great deal of astronomical research depends on surveying large areas of the sky. Large-scale surveys range from searching for pulsars to regions of star formation, from near-Earth asteroids to the distribution of galaxies in the universe. Interesting objects found during surveys can later be targeted for closer scrutiny.

It isn't surprising, then, that astronomers have always felt the need to broaden their horizons and so set about designing telescopes with large fields of view as well as great light-gathering power. One of the most successful range of telescopes is the Schmidt telescope design, which has a wide field of view and produces bright images.

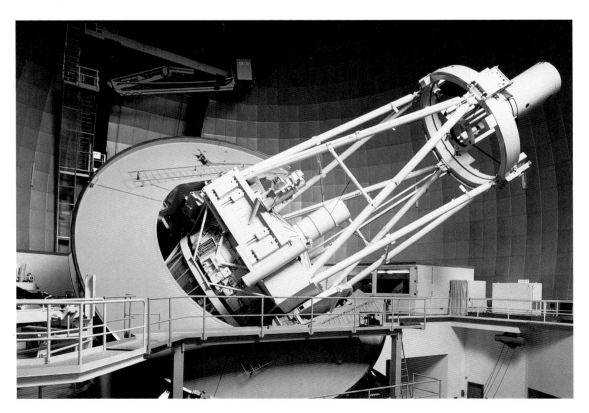

DIFFERENT MOUNTINGS

The 4m telescopes, built up until the 1970s, were to be the last of their kind in one important respect – the way in which they counteracted the Earth's rotation. Until then, telescopes of all sizes had been supported by 'equatorial mounts' whose axes were aligned with the poles of the Earth. This allowed them to track objects across the sky by turning on one axis alone. But making such mountings for larger telescopes became virtually impossible because of the size necessary to support the mirrors: the strength-to-weight ratio was so enormous that the cost was beyond even international financial partnerships.

The alternative was to use an altazimuth mounting, which moves up and down and side to side. By themselves, altazimuth mounts make it awkward to track stars in the sky but advances in computer technology enable astronomers to guide a telescope as precisely as one mounted equatorially.

Sitting closer to the ground, the new generation of super telescopes seemed squat by comparison, but with shorter support structures, the altazimuth design allowed the creation of telescopes with apertures up to 10m, twice the diameter – and four times the light-gathering power – of the Hale telescope.

INTERFEROMETRY AND THE KECK

The largest telescopes in the world at the moment are a pair of 10m instruments called the 'Keck telescopes' Rather than having single 10m mirrors, the Keck telescopes' mirrors are made up of segments.

About 10m seems to be the limit to how big a telescope aperture can be made, but astronomers were not content with such a limitation. Gradually, they learned to combine the light from widely spaced mirrors to simulate the performance of a single large mirror. All the mirrors gathered light and bent it to meet the same focus, acting like a single mirror. In a sense, it is like building the edges of a single huge mirror but not bothering to fill in the 'middle'.

As with a single mirror, the wider apart the edges of the mirror, the better the resolution. The advantages of the technique, called interferometry, return to the two main goals of making larger telescopes: better resolution and better light-gathering power.

One example of an interferometer is called SUSI, the Sydney University Stellar Interferometer in Australia. SUSI consists of widely spaced 15cm-diameter mirrors, simulating a telescope with an aperture of hundreds of metres. The resolution of such a 'telescope' will allow astronomers, among other things, to measure the diameters of stars.

THE VLT

One of the most impressive applications of optical interferometry can be found on a high mountain called Cerro Paranal in the Chilean Andes. Here, the European Southern Observatory is on track to completing the world's largest optical telescope, the Very Large Telescope (VLT).

Its name is too modest: by combining the light of four 8m telescopes standing side by side, the VLT will behave as if it

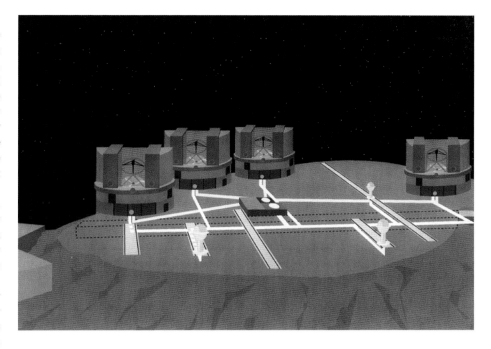

were a single 16m telescope. The construction of a 16m mirror is almost impossible, but by using the interferometry technique a 16m equivalent telescope has become a reality.

However, at present, the most spectacular example is the combination of the twin Keck telescopes. Although these instruments have the light-gathering power of two 10m mirrors (less than the VLT), when used as an interferometer their separation will allow them to simulate the resolution of an 85m telescope. Smaller 'outrigger' telescopes will be added in the future to give the interferometer enough resolving power to search for planets around other stars.

The Very Large Telescope (VLT) is an optical telescope on Cerro Paranal at the European Southern Observatory. VLT1 saw first light in May 1998. The interferometer will be able to resolve objects down to $\frac{1}{1000}$ of an arc second across and the ambitious observing plans include looking for a black hole at the centre of the Milky Way and examining the surfaces of stars.

The 10m Keck I, completed in 1992, is comprised of 36 hexagonal segments, each one 1.8m across. This honeycomb structure means that the mirror is far lighter than if it were in one piece and more manoeuvrable.

Equipment 164–165 ▶

The Mount Wilson Observatory was far from the lights of Los Angeles in 1908 (top), but 80 years later the city had spread so much that observatory was swamped by the city's skyglow (above).

bartender trying to fill a stationary glass, some of the light inevitably spills outside the aperture of any instrument attached to the telescope. With the incredible faintness of some of the objects astronomers observe, every photon counts.

THE SOLUTION: REACH HIGHER

The ideal solution, of course, is to get above the atmosphere and into space. But while this has been done with great success, it is fraught with difficulty and is very, very expensive.

The alternative is a compromise: if you can't get above all of the Earth's atmosphere, at least get above as much of it as you can. This is why modern observatories are built on some of the world's highest mountains. For example, the telescopes on Mauna Kea in Hawaii peer at the universe from a height of 4205m above the surrounding ocean, and the European Southern Observatory at La Silla in Chile is at an altitude of 2400m.

Even at this altitude, the density of the Earth's atmosphere is still significant and still has an effect on the quality of images. To overcome this problem, astronomers have developed a technique called adaptive optics. This is a system of tiltable and deformable mirrors that counteract the effects of the atmosphere. All new observatories are fitted with adaptive optic systems.

THE FIRST HURDLE: POLLUTED SKIES

Astronomers have found that the spread of civilization brings its own interference: light pollution, air pollution and artificial radio transmissions. In the early days of professional astronomy, astronomers had no idea that artificial interference would have such dramatic effects on their work. Their primary concern, when selecting the sites for observatories, was accessibility; many observatories, even this century, were built near the institutions that paid for them. The resulting tragedy is that some of the world's finest and most productive observatories have been seriously compromised by the spread of adjacent cities.

In the second half of the 20th century, astronomers have tried to get away from the pollution, taking them to some of the most inhospitable places on Earth. Since transport isn't the problem it once was, the major consideration in selecting a site for an astronomical observatory is freedom from natural and artificial interference. Avoiding artificial interference is as simple as getting away from populated areas, but avoiding the turbulent effects of the Earth's atmosphere isn't always as easy.

While the stars look sharp and bright to the naked eye, to modern astronomers using state-of-the-art telescopes, observing from sea-level is like standing on the edge of a swimming pool trying to read this book at the bottom of the water. You can make out the big headings, but the really interesting stuff is just a blur.

As well as blurring a star's image, the Earth's atmosphere moves it around in an unpredictable manner. This random motion causes the image to dance around in the field of view of the telescope. Like a nervous

COSMIC RAYS IN THE OUTBACK

For some astronomical pursuits, being at the bottom of an atmospheric ocean isn't too much of a problem. Astronomers who study the most energetic form of radiation known, cosmic rays, are unable to form clear images at the best of times, and so settle for simply being able to detect their elusive prey.

As the highly energetic particles fly in from space, they interact with the air to produce a special type of light called Cerenkov radiation. These fleeting flashes of light are so brief and faint that the human eye is unable to see them, but sensitive detectors placed at the focus of large mirrors are able not only to detect these flashes, but to create crude images of the tell-tale streak of light as the particle enters the atmosphere.

For these cosmic ray astronomers, darkness is the key. One of the darkest places on the Earth is in the Australian outback, although the peace of these astronomers is occasionally broken by kangaroo shooters equipped with spot lights and rifles!

TO ANTARCTICA

The advent of remote observing techniques (observing away from the telescope) where astronomers can either operate a telescope via the Internet or have a dedicated telescope operator do the work for them, means that transport becomes even less of a problem. This has led one group of astronomers to suggest that the best place on Earth to build an observatory is Antarctica. There, surrounded by ice and little else, the conditions for observing are ideal: on a high plateau, the air is still and clear ... and dark for six months at a time.

Antarctica is an ideal place for infrared astronomers who

◀ 36 The Earth's atmosphere

have to battle with the fact that moisture in the atmosphere absorbs infrared radiation. Here it is so dry, all the moisture is frozen solid on the ground. Also, being so cold, the area is free of heat sources that interfere with observations.

THE FINAL FRONTIER

Another place to go, entirely free from atmospheric problems, is into space. High above, in the Earth's orbit, astronomical observatories are literally and metaphorically closer to the stars.

The best known space observatory is the Hubble Space Telescope (HST). This is the first of NASA's planned 'great observatories', and uses a 2.4m mirror to focus pristine, unfiltered starlight into a range of sophisticated instruments. More than any other space-based telescope so far, the HST has provided the most stunning images of the universe humans have ever seen, sparking the imaginations of future scientists and the public. More importantly, HST continues to provide astronomers with scientific data that will keep theorists busy for decades.

The European Southern Observatory is at La Silla, Chile, where the limiting magnitude is +24 or more. At an altitude of 2400m the seeing is exceptional due to the clear, dry and stable climate.

The electromagnetic spectrum 140–141 ▶

Hubble's impact 144–145 ▶

Diminishing darkness 158–159 ▶

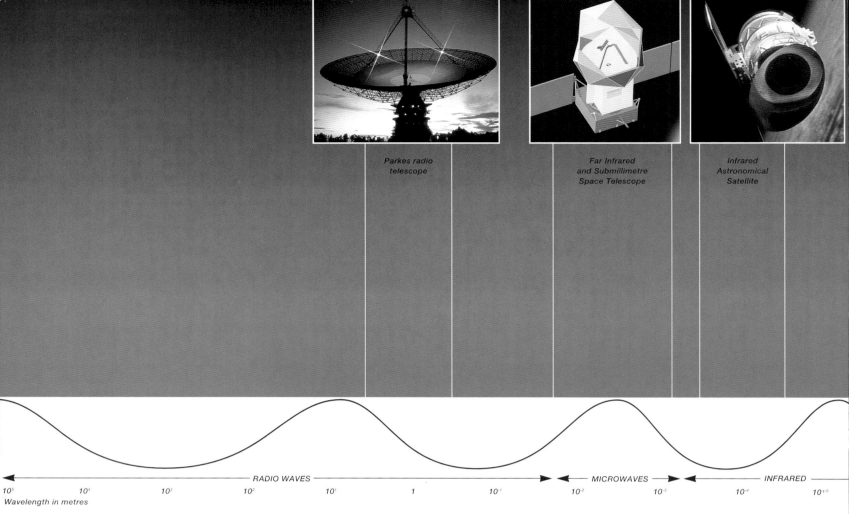

Parkes radio telescope

Far Infrared and Submillimetre Space Telescope

Infrared Astronomical Satellite

RADIO WAVES ———→   ←— MICROWAVES —→ ←— INFRARED —

| 10^5 | 10^4 | 10^3 | 10^2 | 10^1 | 1 | 10^{-1} | 10^{-2} | 10^{-3} | 10^{-4} | 10^{-5} |

Wavelength in metres

THE ELECTROMAGNETIC SPECTRUM

Objects emit radiation across the whole EM spectrum but they can be seen only from the surface of the Earth in visible light and at some radio wavelengths. Radiation at other wavelengths is blocked by the Earth's atmosphere, so to observe in microwaves, infrared, ultraviolet, X-rays and gamma rays astronomers have to use space-borne detectors above the atmosphere.

OTHER WAVELENGTHS

Visible light is just one small part of the electromagnetic spectrum (above). There is nothing special about visible light, it is just the part of the electromagnetic spectrum to which our detectors (our eyes) respond.

The colours we see around us in everyday life represent the way different elements and molecules reflect light. We use this 'spectral' information all the time: when deciding if a banana is ripe or a steak well done. If the different colours of everyday objects can tell us so much with a simple glance in visible light, imagine what we could learn if we were able to see in other wavelengths as well.

OTHER WINDOWS TO THE SKY

The Earth's atmosphere is important to humans, not only for the air we breathe, but because it screens us from harmful radiation. However, this means that it is impossible to see the universe at some wavelengths because the Earth's atmosphere is not transparent enough.

One of these 'atmospheric windows' is, obviously, in optical wavelengths (visible light). Another type of radiation that makes it through the atmosphere is radio. The radio sky appears to be very different from the visible sky we're used to. Gone are the familiar stars, replaced by new patterns of radio-bright celestial objects. The temptation to explore the new sky was overwhelming.

THE RADIO SKY

To explore the radio sky, radio astronomers need larger apertures for the same reasons as their optical colleagues:

to collect more radiation to gain more information. But while optical astronomers have it tough, radio astronomers have it tougher. Resolution in the radio band of the spectrum is more difficult than in the optical because of the longer wavelengths involved.

Despite an increase in size, the resolving power of the much larger radio dishes is often inferior to that of optical telescopes. For example, the largest radio telescope in the southern hemisphere, the 64m Parkes radio telescope in New South Wales, Australia, has an angular resolution of three minutes of arc, or 0.05°. This is equivalent to seeing a car at the distance of nearly 2km. The much smaller 3.9m Anglo-Australian Telescope also in New South Wales, which works in the optical and infrared, can see details 180 times finer, ie, you would be able to read the car's number plate.

This is the reason why radio telescopes dwarf their optical counterparts physically. While the largest optical telescopes are the 10m Keck telescopes, the largest radio telescope in the world is the 300m radio dish built into a valley in Puerto Rico.

Radio astronomers applied the principle of interferometry to their own field. At first, radio interferometers were measured in tens of metres, then hundreds, then thousands. By the time the smaller dishes were separated by kilometres the technique was dubbed 'very long baseline interferometry', or VLBI. When spread over distances that span continents, VLBI gives radio astronomers a sharpness of vision exceeding that possible with 4m optical telescopes.

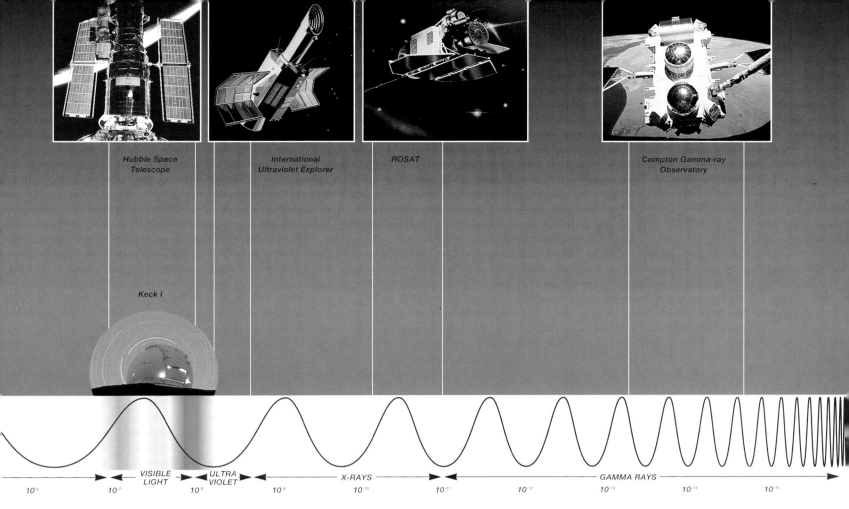

Hubble Space Telescope

International Ultraviolet Explorer

ROSAT

Compton Gamma-ray Observatory

Keck I

VISIBLE LIGHT	ULTRA VIOLET		X-RAYS			GAMMA RAYS				
10^{-6}	10^{-7}	10^{-8}	10^{-9}	10^{-10}	10^{-11}	10^{-12}	10^{-13}	10^{-14}	10^{-15}

Not content with continent-sized radio telescopes, VLBI was taken to new heights, literally. Using a technique called 'space VLBI', a small radio dish was placed into Earth orbit and linked with dishes on the ground. The Space-Earth interferometer provides the resolution of a radio telescope tens of thousands of kilometres in diameter, larger than the Earth itself and this means that astronomers can detect radio waves at greater distances than ever before.

THE INFRARED SKY

The edges of the atmospheric windows are not sharply defined and on either side of the optical window, astronomers can see a little into the realm of invisible radiation (above).

Infrared radiation was discovered in 1800 and is just beyond the red end of the visible spectrum. Infrared radiation is often called 'heat' radiation because of the sensation it produces in living tissue. For example, you can feel infrared radiation coming from a stove element when you turn it on long before it begins to glow dull red in the visible.

To the astronomer, however, infrared radiation reveals relatively cool phenomena. In comparison with other types of radiation, infrared is emitted by relatively peaceful processes in the cosmos, such as occur in disks of dust surrounding newly formed stars. But infrared has another important property: it can penetrate clouds of dust that obscure astronomers' views at shorter wavelengths, including visible light.

This allows astronomers to peer through obscuring interstellar clouds at events going on within the clouds, such as the birth of stars.

The HST is not the only successful space observatory. Producing equally valuable images and data, but in the infrared, is the Infrared Space Observatory (ISO). With an array of sophisticated infrared detectors, ISO gazes at the infrared universe using detectors with a resolution 10 times sharper than anything flown before, and a thousand times more sensitive. If it had to, ISO could detect the warmth of a human being in space at a distance of 100km.

THE UV, X-RAY AND GAMMA RAY SKY

A year after infrared radiation was discovered, ultraviolet radiation was detected in sunlight. Ultraviolet radiation is beyond the blue part of the visible spectrum.

Towards the end of the 1800s, ultraviolet spectra from stars showed that the stars were abundant in hydrogen. Viewing the ultraviolet sky showed astronomers that hydrogen is the most abundant element in the universe. This is an example of how important it is to study the sky in as many different wavelengths as possible: there are many celestial treasures visible only at certain 'colours'.

Another realm of the electromagnetic spectrum that had to wait for the space age is X-rays. Space X-ray observatories like the Einstein Observatory and ROSAT have been spectacularly successful.

Observatories have now been sent aloft to observe at many wavelengths, among them the International Ultraviolet Explorer and the Gamma Ray Observatory.

LEFT TO RIGHT (ABOVE)
Radio waves: The Parkes radio telescope observes between 100mHz–45gHz.
Microwaves: The Far Infrared and Submillimetre Space Telescope (FIRST), to be launched in the in the 21st century, will observe in wavelengths between 100–1mm.
Infrared: The Infrared Astronomical Satellite (IRAS) made an all sky survey between wavelengths 10–100mm.
Visible light: The Hubble Space Telescope and Keck I, as well as other ground-based telescopes, operate in this wavelength.
Ultraviolet: The International Ultraviolet Explorer (IUE) operates in the ranges 115–190mm and 180–320mm.
X-rays: The Röntgenstraheln Satellite (ROSAT) made a deep all sky survey in the 0.2–3keV band.
Gamma rays: The Compton Gamma-ray Observatory (CGRO) observes in the 15keV to 30GeV part of the EM spectrum.

Spectroscopy 142 ►

Hubble's impact 144–145 ►

RECORDING THE STARS

No matter how good the telescope, some method of recording what is seen is necessary. For two-and-a-half centuries following the invention of the telescope, the human eye was the primary detector and the human hand was the primary recorder. In the mid-19th century, however, photography emerged as the first modern method by which astronomers could make a permanent, unbiased record of celestial phenomena. The photographic revolution changed the way in which astronomy was done, and despite fantastic and invaluable advances in electronic imaging, it remains one of the most important tools in astronomical research.

Photography also revolutionized the cataloguing of stars. A fundamental property of stars and galaxies is their brightness, and determining brightness is called photometry. The size and density of a star's photographic image is related to its brightness, and so it is possible to determine brightness accurately from photographs.

Photographs reveal fainter and, therefore, more stars than are visible to the human eye through a telescope, but they also make a permanent record of their relative positions that can be measured and checked. This made the process of cataloguing the stars much faster and easier than ever before. It also provided an historical record of the appearance of the sky, and of individual stars, which could be used when identifying the precursor of a new object such as a supernova. A famous example of this is when Robert McNaught found the precursor to the supernova SN1987a on photographs at the Anglo-Australian Observatory.

SNAPSHOTS OF THE SOLAR SYSTEM

Photography also changed the way the solar system was studied. The first asteroid was discovered in 1801 and by 1891, when an asteroid was discovered photographically for the first time, there were only 300 known. Suddenly asteroids were found everywhere: within 50 years the number of known asteroids had risen to 1500.

Photographs of the major planets, in particular the ever-changing face of Jupiter, also grew as an important astronomical pursuit. Although photographic resolution is inferior to that possible with the human eye, the permanency and hence reliability of planetary photography changed the way astronomers study the planets.

SPECTROSCOPY

One of the major contributions of photography was the permanent recording of the spectra of stars and other celestial objects. Newton had shown how white light, when passed through a prism, dispersed into its component colours, or spectrum. In the spectrum of the Sun – and, as was soon discovered, any other star – can be seen a series of dark lines. These are absorption and emission lines of gases in the star's atmosphere and of matter that lies between the Earth and the stars. These were noticed in 1802, but they were first studied seriously by the German physicist Joseph von Fraunhofer in 1814. These 'Fraunhofer lines' reveal the elements present and their discovery led to the science of spectroscopy.

Spectroscopes were soon attached to telescopes and by 1863, William Huggins was able to publish lists of the spectral lines in stars. A further major advance was made in 1868 when Huggins showed how the motions of stars could be found by measuring a phenomenon called Doppler shift. The Doppler shift means that astronomers can use spectroscopy to determine the speeds with which stars – and anything else in the sky – moves towards or away from the Earth.

ELECTRONIC IMAGES

Despite the continuing success and use of photography, the most efficient way of producing images of celestial objects today is by using some form of electronic detector. CCDs (charge-coupled devices) are the most common detector used, although their exact construction and make-up varies depending on the type of work for which they are intended.

CCDs are essentially electronic film; while ordinary film needs hours of exposure to create an image, CCDs can achieve comparable results in a matter of minutes. This gives astronomers instant feedback on the success or otherwise of their efforts to take an image of a celestial object.

As electronic images of bright objects can be created in shorter periods of time than on film, it follows that

The Earth has her own ring system, made of man-made satellites in geostationary and other orbits.

electronic detectors are capable of revealing much fainter objects than film over longer periods. So sensitive are modern detectors, in fact, that some are now capable of sensing individual photons, the basic particles of light.

A DANGEROUS ENVIRONMENT

A major concern facing all space users is space debris. As more and more satellites are being sent into the Earth's orbit – and left there at the end of their life – the near-Earth environs of space are becoming incredibly crowded. The gradual accumulation of these mechanical corpses has taken a surprisingly small time to become a serious problem. In less than a century, the loss of an expensive satellite or observatory being struck by a piece of flying debris has become a significant risk. This is more than just an academic problem: it is suspected that more than one scientific experiment or communications satellite has already been disabled, damaged or destroyed by collision with space debris.

Even if space debris were not a problem, there is no guarantee that a telescope will last long in the harsh environment of space. Radiation from the Sun and other sources attacks spacecraft continually. This is why the Hubble Space Telescope was given a range of Key Projects: special investigations that were given the highest priority to stand a better chance of completion before the telescope experiences its inevitable demise.

NO PERFECT SOLUTION

The stunning success of observatories like the Hubble Space Telescope suggests space may be the place to go, but the cost of building a telescope like Hubble, putting it in orbit and maintaining it are enormous. Even if the funds to build more of these orbiting observatories could be found, astronomers like to swap and change the instruments attached to their telescopes on a regular, sometimes nightly basis. This is something that just can't be done with space-based telescopes; their complement of instruments can only be changed during major servicing missions, which, if they occur at all, are years apart.

There is no perfect solution to studying the universe, except to look in as many ways as possible. Over the 400 years since Galileo and his contemporaries turned their tiny telescopes onto the sky, discovery has followed discovery, each one as unpredictable as the one before. Each time a new band of the electromagnetic spectrum is opened, each time a more sensitive detector is made, it is like turning a toy over again, and another facet of the universe turns into view in the hands of the astronomers.

Hubble's impact 144–145 ►

Astrophotography 165 ►

CCDs 165 ►

Cataloguing the stars 166 ►

HUBBLE'S IMPACT

Above every square metre of Earth's surface lie 10 tonnes of often cloudy, humid and polluted air. Ground-based telescopes have allowed us to peer through this fog, but it was not until the space age that we could get above the atmosphere.

BY DR MIKE DISNEY

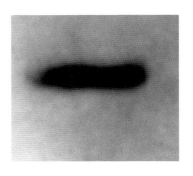

The Hubble Space Telescope imaged small black dots in the Orion nebula 1500 light years away. When observed closely, they were seen to be disk-like structures of cold gas and dust surrounding young stars, possibly solar systems in the making.

" Orbiting freely above the atmosphere, the Hubble Space Telescope (HST) can spy on the cosmos with something like 1000 times as much detail as an Earth-bound telescope – with really enlightening results. Stars that appear like large, wobbly jellyfish when seen through our atmosphere, appear to the HST as sharp points of light. This enables the repaired HST to detect them as much as 10 times farther away. The HST can also analyse ultraviolet radiation, which often contains vital spectroscopic data – hot, young stars are often more easily visible in ultraviolet, for example – but this radiation is absorbed by Earth's atmosphere before it reaches the ground.

Such is the HST's revolutionary performance that there is hardly a professional astronomer on Earth who is not working either directly on HST data or on the consequences that such data generates. Even in fields where we were told the HST would have little impact, such as observations of faint galaxies, it is still shaking the subject to the core. When, a few years ago, the HST undertook its revolutionary survey of a distant region of space, now known as the Hubble Deep Field, it gave us the first sight of more than 3000 distant galaxies. Scientists believe that some of these galaxies are so young and so far away that we are actually seeing them being born out of pristine gas. Young astronomers are so intrigued by pictures such as these that they are being drawn in droves into the study of galaxy formation, and large ground-based telescopes are being hastily re-programmed to follow up the HST's unexpected insights.

The HST has even shed light on what is perhaps the most challenging problem facing astrophysicists to date: the nature of the dark matter that binds galaxies together. Galaxies spin too fast to be held together by the gravitational effects of the stars we can see in them; so, some massive, dim component must be lurking there, too. The most conservative explanation for this hidden matter is that galaxies possess an overwhelming population of very faint, red dwarf stars. However, just one HST deep-space image of the halo of our own Milky Way was enough to

The dinosaurs vanished 65 million years ago, but the HST can see galaxies as they were 10,000 million years ago.

challenge this argument, because the HST did not see dwarfs in anything like the required numbers. So, we had to seek imaginative alternatives, such as exotic subatomic particles. One thing is certain: something mysterious dominates our cosmos.

Quasars, the most distant, most luminous objects in the universe, are so far away that when seen with ground-based telescopes, they appear as mere points of light with a faint, mostly indistinct fuzz around them. Using the HST, however, we can see clearly that they are the nuclei of spiral and elliptical galaxies, which are usually (but not always) undergoing collisions or interactions. Astrophysicists think that at least one of these strange interacting galaxies contains a black hole weighing as much as a thousand million Suns, and that debris from galactic interactions falls into that black hole, rendering up enough of its energy to feed the quasar's enormous output. This also explains the otherwise surprising evolution of the quasar population. They were very rare in the early universe because galaxies didn't then exist. They are very rare today because the expansion of the universe has pulled galaxies far apart. But some 10 billion years ago, when galaxies were more common and closer together, quasars were a thousand times more common than they are today.

This research illustrates another powerful attribute of HST: its value as a time machine. Since light has a finite speed, when we look out in space, we are looking backward in time, and so any images we see of distant galaxies are actually made up of light that left the galaxies long ago. The dinosaurs may have vanished 65 million years ago, but the HST can see galaxies as they were 10,000 million years ago. By using the HST, we can see the universe as it was in the distant past and compare it with the universe seen today – all of which tells us more about how the universe itself actually began, and makes us better able to predict where it is heading.

We don't have to look far away to plumb the universe's ancient past. The sheer precision of HST measurements of the colours of stars in our own galaxy and its neighbours is enabling us to

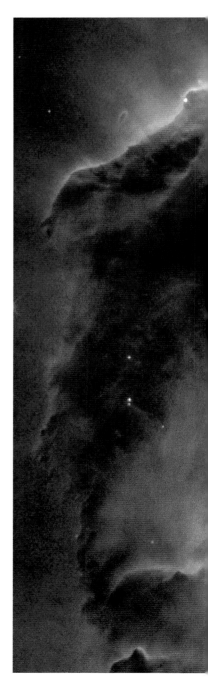

decode the history of star formation within them – a sort of archaeology on a cosmic scale.

How do stars and planets originate? HST images of young stars in Orion show most of them surrounded by large disks of cold gas and smoke. It is likely that planets will form out of these disks, which means that planets are not likely to be rare. On the contrary, HST findings suggest that they are probably the rule rather than exception – in which case there may be billions of Earth-like planets in our own galaxy alone.

In the long run, it is the philosophical impact of a scientific discovery that counts the most. Galileo and his telescope removed humankind from the centre of the cosmos. Darwin demoted us to a transient stage in evolution's pitiless game. Where will the HST lead us? So far, we haven't had to face such a shocking HST discovery, but it is very early days.

What is most exciting to me about HST is that it is freely sharing exciting new data with anyone around the world who has the curiosity to take part in the excitement. And about time, too! The danger of Big Science is its tendency to form a church with a hierarchy of high priests. NASA's open policy regarding the HST is to be applauded. It is my hope that the first shocking discovery with the HST will come from some young Galileo not too pickled in the establishment view. If that happens, then we will see science move back a little towards where it really belongs: in the mainstream of human endeavour. **"**

Hubble revealed the structure of the Eagle nebula in incredible detail, showing tall pillars of dust and gas. This cool interstellar dust and gas is an incubator for new stars. As the dust and gas are slowly boiled away by ultraviolet radiation from nearby stars, globules of denser gas are revealed within which embryo stars are forming. (The blacked-out area in the top right was not imaged.)

SPACEFLIGHT HISTORY

ROCKETS MOVE FORWARDS BY EJECTING PROPELLANT
BACKWARDS AT VERY HIGH SPEED. THEY PUSH AND
REST ON NOTHING, SO THEY CAN WORK IN SPACE.

As with so many advances in technology, rockets were developed for military purposes.

The invention and development of rockets have made it possible for Mankind to work and live in space, and have opened up the heavens for astronomers to study.

EARLY BREAKTHROUGHS

The earliest rockets were invented in China before the 13th century and were propelled by burning gunpowder. Between the years of 1895 and 1923, Konstantin Tsiolkovsky and Hermann Oberth invented rocket-propelled space ships, which were fuelled by liquid fuels. Hermann Oberth foresaw the developments of the space age, predicting that scientists would be able, in time, to build a space station, from which they would have the opportunity to observe the Earth and relay radio messages, and which would act as a base for further exploration.

Rocketry became practical during the Second World War, when engineers such as Sergei Korolev in Russia and Wernher von Braun in Germany developed rockets for military purposes.

In 1945, Arthur C Clarke identified geostationary orbits, which have orbital periods of 24 hours. These allow telecommunication and weather satellites to stay over the same place on the Earth. This technique was used for the first time to televise the 1964 Olympic Games.

THE SPACE AGE BEGINS

The space age started in 1957 with the launch of the first artificial satellite, Sputnik 1, which orbited the Earth and carried simple instruments.

At present, the main use of space is for worldwide transmission of TV, telephone and computer data; the largest telecommunications satellites weigh 7 tonnes. Earth observation satellites are the next most frequent category, for weather forecasting and resource exploration. The biggest growth sector is expected to be in satellites like the Global Positioning System (GPS) by which ships, aeroplanes and cars can navigate.

MANNED SPACE EXPLORATION

The first man to go into space was Yuri Gagarin who orbited the Earth in 1961. Man first walked in space in 1965, and visited the Moon in 1969 when Neil Armstrong stepped from Apollo 11 and took his giant leap for Mankind.

Oberth's vision became real with the construction in 1986 of the Mir space station. The International Space Station will be the next step.

The next human steps into space may be to return to the Moon to establish a lunar colony, or to explore Mars.

FUTURE ROBOTIC EXPLORATION

Many scientists advocate exploration missions by robotic spacecraft, looking down to study the Earth as a planet, looking up to study the universe and exploring the solar system by taking equipment to distant places.

Economic considerations have turned both NASA and ESA away from the giant scientific observatories toward a 'smaller, faster, cheaper, better' philosophy of imaginative, smaller, quickly produced, well-targeted space missions.

Right: Exploring space is expensive and new technology, for instance the International Space Station, is now being developed through cooperation between countries rather than competition.

Far right: Rockets, such as the European Space Agency's Ariane, with some reusable components are still a primary method for putting satellites into orbit.

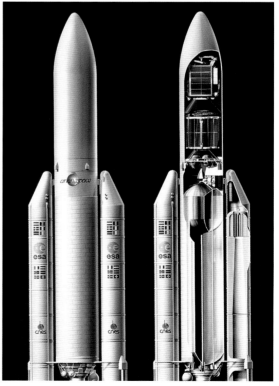

SPACE MILESTONES

Date	Name	Achievements
1926 Mar 16		First launch of liquid fuel rocket (Robert Goddard).
1942 Aug 16	A4	First rocket launch to break through the sound barrier (V2 prototype).
1957 Oct 4	Sputnik 1	First satellite launched into space.
1961 Apr 12	Vostok 1	The first manned space flight (Yuri Gagarin).
1963 Jun 16	Vostok 6	First woman to go into space (Valentina Tereshkova).
1963 Jul 26	Syncom 2	First geostationary satellite.
1965 Mar 18	Voshkod	First space walk (Aleksei Leonov).
1969 Jul 21	Apollo 11	First men to walk on the Moon (Neil Armstrong, Buzz Aldrin).
1981 Apr 12	Columbia	First flight of re-usable space shuttle.
1983 Jun 13	Pioneer 10	First space vehicle to move outside the planets.
1986 Feb 20	Mir	Modern generation space station.

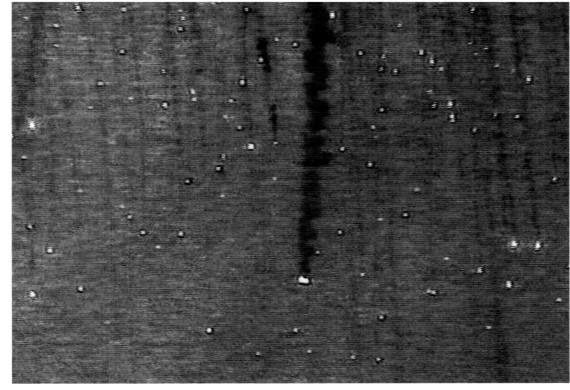

Above: NASA's fleet of shuttles provides both a means of getting hardware into orbit and a platform for astronauts to work on.

Left: Space can be a dangerous place, and mistakes can prove expensive, as when Ariane 5's guidance systems failed and the rocket had to be destroyed.

ROBOTIC MISSIONS

THE COSTS AND DANGERS OF MANNED SPACEFLIGHT ARE SO GREAT THAT MANY SPACE SCIENTISTS ADVOCATE 'GO THERE' MISSIONS BY ROBOTIC SPACECRAFT.

Astronomers have been trying to discover more about the universe for centuries. The first major discovery in space science was the discovery of the Van Allen belts of radiation around the Earth by Explorer 1 in 1958.

THE MOON

Remotely operated landers (Lunakhod) studied the Moon in the 1960s and the Apollo astronauts collected 380kg of lunar rock and brought it back to Earth for analysis. In addition, missions like Clementine and the Lunar Prospector have revealed even more about our nearest neighbour.

ORBITERS

Orbiters go into orbit around planets examining their atmospheres and sometimes mapping out their surfaces. The Mariner series of orbiters were the first to visit the nearest planets to Earth. The two Voyager spacecraft launched in 1977 made a 'grand tour' of the major planets, revealing their amazing satellites. Magellan, a radar-imaging spacecraft, surveyed Venus' volcanic surface, Giotto captured images of fountains of melting ice and dust from comet Halley, and Galileo measured Jupiter's behaviour. Cassini, a comprehensive space observatory to Saturn, will be followed by Rosetta to comet Wirtanen and the ESA Mission to Mercury. Pluto is the only planet still to be explored.

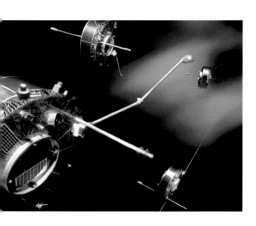

Above: The Cluster satellites were designed to monitor the Sun from above the Earth's atmosphere and to provide data on the effects of solar events and the solar wind on the near-Earth environment.

Right: The Galileo probe entered Jupiter's atmosphere in December 1995 and spent nearly an hour taking measurements before the massive pressure crushed it.

LANDERS

In addition to the Moon, landers have travelled successfully to the surfaces of Venus and Mars. The Russian Venera probes did not last long in Venus' inhospitable environment, but the Viking and Pathfinder missions closely examined regions on the surface of Mars.

The Galileo spacecraft to Jupiter launched a probe into the atmosphere while the Cassini mission will send the Huygens probe into the thick atmosphere of Saturn's largest moon, Titan.

X-RAY ASTRONOMY

Astronomers quickly spotted the potential of spaceflight to observe cosmic radiation, blocked by the Earth's atmosphere. Riccardo Giacconi and H Gursky launched a suborbital rocket with the intention of examining the composition of the Moon by looking at X-rays that fluoresced from its surface under the stimulus of the Sun's radiation. The brief rocket flight failed in its primary objective, but did see a source of celestial X-rays beyond the Moon in the constellation of Scorpius. This was the start of X-ray astronomy.

ASTRONOMY FROM SPACE

There is an imaginative range of past and future 'look-up' missions, which allow astronomers to survey the universe in the entire electromagnetic spectrum. Possibly the most influential space observatory is the Hubble Space Telescope, operating in the optical spectral band accessible from the ground but only through the blurry atmosphere.

THE NEXT FRONTIER

One great area still to be explored is gravitational waves. They are undetected as yet but their existence is firmly predicted by Einstein's Theory of General Relativity and by the slowing of the orbit of a binary pulsar, which is affected by gravitational waves. Plans for LISA, a 50 million kilometre free-flying interferometer to detect objects emitting gravitational waves, are being developed.

EXPLORATION OF THE SOLAR SYSTEM

Destination/Date	Name	Achievements
Moon		
1969 Jul 21	Apollo 11	First men on the Moon.
Mercury		
1973 Nov 4	Mariner 10	First Mercury fly-by.
Venus		
1970 Aug 17	Venera 7	First Venus landing.
Mars		
1964 Nov 28	Mariner 4	First Mars fly-by.
1975 Aug 20, Sep 5	Viking 1 and 2	First Landers, surface images, soil analysis.
Jupiter		
1972 Mar 2	Pioneer 10	First Jupiter fly-by.
1977 Aug 20, Sep 5	Voyagers 2 and 1	Multiple fly-bys, exploration of satellites.
1989 Oct 18	Galileo	First atmospheric probe.
Saturn		
1973 Apr 5	Pioneer 11	First Saturn fly-by.
1977 Aug 20, Sep 5	Voyager 2 and 1	Multiple fly-bys.
Uranus, Neptune		
1977 Aug 20	Voyager 2	First fly-bys.
Sun		
1990 Oct 6	Ulysses	First orbit over the poles.
Comets, asteroids		
1985 Jul 2	Giotto	Mission to comet Halley.
1989 Oct 18	Galileo	First asteroid encounter.

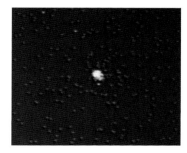

Above: To observe at X-ray wavelengths, astronomers have to use spacecraft above the Earth's atmosphere. These 'before, during and after' images of an X-ray burster could not have been obtained with any ground-based instrument.

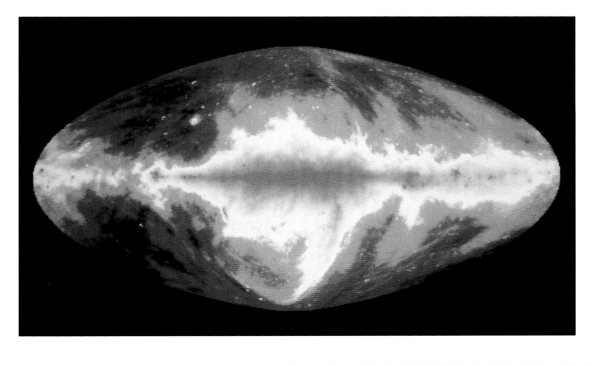

Left: Not all radio wavelengths can get through the blanket of the Earth's atmosphere, so better views are obtained from orbiting spacecraft. This image shows the Milky Way at radio wavelengths.

LOOKING FOR ET

The idea that there might be other beings somewhere in the universe is an old one, but proving it is quite another thing. Only in the last 50 years of the 20th century has it been possible to make a scientific search for intelligence elsewhere.

BY DR SETH SHOSTAK

Above: The white line stretching across this computer screen is the signal from the Voyager 2 spacecraft at a distance of about four billion miles. SETI scientists used Voyager's signal to check their system.

" One way to look for extraterrestrials would be to go to other planets. However, although it is feasible to build spacecraft to go to the Moon and other planets of our solar system, none of these nearby worlds will have the sort of creatures we seek: intelligent, interesting beings.

So we must search around other stars. Sadly, our current rockets can't take us to such distant realms. Even the nearest star, Proxima Centauri, is 10,000 times farther away than Pluto. A rocket would take 80,000 years: one-way.

It is also unlikely that any extraterrestrials will come here. Despite the popular idea that UFOs are alien spacecraft buzzing the countryside, few scientists are convinced that these lights in the sky have anything to do with extraterrestrials. All of which begs the question: how can we find any extraterrestrials, if they do exist?

By the end of the 1950s, researchers realized that radio is an effective way to send information between star systems. Radio needs relatively little power to broadcast a signal that could be picked up light years away. The energy cost of sending a radio message to the nearest stars is only about one US dollar a word; far lower than trying to go there. And radio messages move at the speed of light.

However, even though radio is a great way to get in touch, researchers don't broadcast a hailing signal to the stars hoping for answers. After all, even if there were aliens on a planet orbiting Proxima Centauri, it would take a little over four years for our signal to reach them, and the same length of time for their reply (if they did reply) to get back to Earth. That's too long to wait, so scientists hope that some advanced aliens have been broadcasting for a while, sending signals that are reaching us now.

Researchers from SETI, the Search for Extraterrestrial Intelligence, are using mammoth radio telescopes to hunt for these signals. Searching for extraterrestrial signals is nothing new: the first modern SETI observations were made by astronomer Frank Drake in 1960, using a 26m diameter radio telescope in Green Bank, West Virginia. In Project Ozma, Drake spent

several weeks pointing the telescope towards two nearby Sun-like stars: Tau Ceti and epsilon Eridani. He chose these stars because astronomers believe that these are the most likely to have orbiting planets similar to the Earth. There are also plenty of Sun-like stars around: in our galaxy alone, there are about 50 billion.

The largest modern SETI experiment is Project Phoenix, run by the SETI Institute in Mountain View, California. Whereas Project Ozma searched for alien signals from the vicinities of two nearby stars, Phoenix's target list includes more than a thousand. Drake listened to only one channel at a time; Phoenix listens to tens of millions.

After 40 years of SETI experiments, no-one has heard a convincing alien signal. Does this mean that no-one's out there? Could Earth be the only place in all the vast starfields of our galaxy where creatures who can build radio equipment exist? If so, this would be remarkable. From an astronomical point of view, we don't believe our solar system is that special. So why should we be alone?

The most probable reason why we haven't heard the extraterrestrials yet is that we've only just started to hunt. A search of a thousand stars in our galaxy is akin to sampling a teaspoon of sand out of a large sand-filled living room, where each grain is a star.

In addition, we could be looking at the wrong time or at the wrong spot on the dial. Researchers are listening in the microwave region of the radio band because the universe is naturally quiet at these frequencies – there's less natural background radio static. We know this, and advanced aliens will know it too. Since it's hard to be heard when interference is high (as anyone who's tried to chat next to a waterfall knows), we assume that the extraterrestrials will be clever enough to tune their transmitters to a microwave channel. But again, we could be wrong. Or maybe we're right, but haven't checked out the correct channel yet. Even the most ambitious SETI projects only monitor a small slice of the microwave band.

So when you consider the many uncertainties that are involved in tracking extraterrestrials down, it's not too surprising that we haven't yet

> *After 40 years of SETI experiments, no-one has heard a convincing alien signal. Does this mean that no-one's out there?*

picked up their broadcasts, even if our whole galaxy is teeming with intelligent beings.

A deliberate signal from extraterrestrials would be an astounding discovery. Its very existence would tell us, suddenly and forever, that we are not intellectually or culturally unique. Any aliens that we hear will be at least as technologically advanced as we are, otherwise they won't be broadcasting. Indeed, we can expect them to be thousands or millions of years beyond us.

So, if there is a 'message' with their signal, it might be as hard for us to understand as the output of a computer modem would be for a caveman. Only if the extraterrestrials are intentionally targeting newly emerging societies (and that's what we are, from a cosmic perspective) will they make their message understandable.

Such possibilities are as speculative as they are truly mind-boggling. But the possibility of a SETI success is very real. As digital electronics continue to advance, the receivers used by those who listen for signals from other worlds also improve. Today's experiments are a 100 trillion times more effective than Project Ozma.

And even if we don't actually understand the extraterrestrials, just hearing the faint whispers of a distant civilization would greatly change the way we view our own. **""**

Project Phoenix used the Parkes 64m Radio Antenna situated in New South Wales during the first half of 1995 to scrutinize approximately 200 Sun-like stars for signs of intelligent signals. No signal has yet been found.

THE NIGHT SKY

The night sky is an area of outstanding natural beauty, full of wonders and free to anyone who wants to look up and enjoy its treasures. Even from among city lights, you can see and enjoy the changing face of the Moon.

Away from city lights, the majesty of the Milky Way can be seen. Turning binoculars onto this tenuous band reveals hundreds of stars, and telescopes reveal wispy nebulae and other galaxies, light years away.

The sky changes throughout the year as the Earth orbits the Sun with the patterns of different constellations wheeling across. Among the stars wander the members of our solar system: the planets, comets and asteroids. Some solar system dust can fall to Earth, burning up in the atmosphere as meteors or 'shooting stars'. Every clear night there is something to discover.

MAPPING THE SKY

IMAGINE YOU ARE STANDING BENEATH A HUGE
PLANETARIUM DOME ON WHICH ALL THE STARS,
GALAXIES AND PLANETS ARE PROJECTED.

There is much change and movement in the skies above our planet, yet to look up is to appreciate a fundamental continuity and order. The occasional exotic and unexpected event that hits the headlines every now and again, such as the sudden arrival of a comet from the outskirts of the solar system, is the treat. But the routine outlook is no less wonderful.

FINDING YOUR BEARINGS IN THE SKY

The rising and setting time of the Sun, and its apparent path through the sky, change over the year, as does the length of days and nights. This occurs because Earth has an axial tilt, and different parts of our planet lean towards or away from the Sun in the course of its annual journey.

It is, however, possible to find your way around the night sky. Imagine the Earth's poles and equator extended into space. As the Earth spins, the sky turns about the celestial poles. In the northern hemisphere, this point is marked by the reasonably bright star, Polaris, but south of the equator there is no such convenient indicator.

Depending on your latitude, some stars rise and set in the course of 24 hours, whereas others remain above the horizon, circling the celestial pole. Stars that are always visible are circumpolar stars and can never be seen in the opposite hemisphere. In the north they include Cassiopeia and the pattern of the Plough, or Big Dipper, in Ursa Major. Southern observers have the Southern Cross.

The constellation Orion, the Hunter, is a major signpost of Earth's skies. It is visible from the northern hemisphere at some time during the night for nearly six months. It also dominates the southern hemisphere's summer skies, though there it is seen the other way up. Orion's star-studded belt lies on the celestial equator. Northern observers face south to see him, whereas in the southern hemisphere he commands the sky to the north. At the poles, the celestial equator sits on the horizon, so only half of Orion is visible – a different half from each location, of course!

Similarly, Pegasus, the winged horse of Greek mythology, flies 'upside-down' through the autumn skies of the northern hemisphere, but right way up in the southern spring.

As the Earth rotates completely on its axis once every 24 hours, the stars will appear to rise and set during the night. Stars will leave a trail of light if a camera is pointed at the sky and the shutter left open. This photograph shows star trails around the south celestial pole, which is in the centre of the picture.

◄ 37 Season's and the Earth's orbit

◄ 78–79, 82–85 Comets

To someone at the equator, the celestial poles sit on the horizon, the celestial equator is overhead – and so is Orion. The stars of both hemispheres rise and set every 24 hours. Over a year, all the stars can be seen. Days and nights arrive suddenly, and are always equal in length.

In contrast, Earth's polar regions have months of darkness when the Sun never rises, followed by months of daylight. During polar night, the stars of an entire hemisphere are seen above the horizon.

RIGHT ASCENSION

In describing the positions of objects in the sky, astronomers find it convenient to go back to the ancient notion of a sphere on whose inner surface we see the stars, nebulae, galaxies and planets projected, almost as if it were an immense planetarium dome. Positions on the celestial sphere's underside are defined, just like those on a terrestrial globe, in terms of longitude and latitude.

Celestial longitude is described as Right Ascension (RA), increasing eastwards from a prime meridian. A meridian is a great circle which passes through both poles of the celestial sphere, at right angles to the equator.

RA can be expressed either in degrees (0–360) or, more commonly, in hours, minutes and seconds. A complete rotation of the Earth, taking 24 hours, is equal to 360° of RA. Each hour of RA comprises 15° meaning that every part of the celestial sphere rotates 15° in one hour. A degree consists of 60 minutes, and a minute of 60 seconds. The last two are angular measurements, referring to the arc which they span – hence they are often referred to as arc minutes or arc seconds.

Lines of RA are 'great circles' – they pass through the centre of the celestial sphere, marking out its circumference. The great circles sweep across the sky at right angles to the celestial equator, passing through both poles.

DECLINATION

Lines at right angles to RA are lines of Declination, which mark out the latitude of any object in the sky.

Only one line of latitude in the sky is a great circle: the celestial equator. The celestial equator is a projection onto the sky of the Earth's equator, and has a Declination of zero.

Positions in Declination northwards from the celestial equator are shown as '+', those to the south as '–'. Commonly abbreviated to 'Dec', Declination is measured in degrees, minutes and seconds ranging from 0° at the celestial equator to 90° at the north and south pole.

As an example of the use of RA and Dec, Betelgeuse, the bright red star marking Orion's left shoulder, lies at RA 05h 55m 10s, Dec +07° 24m 25s.

Above left: Part of the constellation of Orion seen from the northern hemisphere. The red supergiant, Betelgeuse, is at the top, while the 'sword' points downwards.

Above right: The constellation of Orion as seen from the southern hemisphere. Betelgeuse is the distinctive red star in the middle near the horizon. From the southern hemisphere, Orion's 'sword' points upwards.

RA AND DEC
The position of celestial objects can be denoted by a longitude and latitude system based on the celestial equator and the First Point of Aries. Right Ascension marks out celestial longitude while Declination indicates latitude.

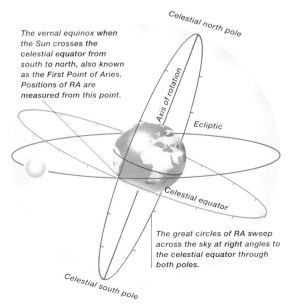

The vernal equinox when the Sun crosses the celestial equator from south to north, also known as the First Point of Aries. Positions of RA are measured from this point.

Celestial north pole

Axis of rotation

Ecliptic

Celestial equator

The great circles of RA sweep across the sky at right angles to the celestial equator through both poles.

Celestial south pole

Observing comets 161–162 ▶

Orion 168–169 ▶

THE ECLIPTIC

A further important great circle on the sky is the ecliptic. Each year the Earth orbits the Sun, but from the Earth it appears that the Sun goes round the Earth. The Sun's apparent path against the background stars is called the ecliptic.

As the Earth's axis is tilted by 23.5° relative to its orbital plane, so the ecliptic is also tilted with respect to the celestial equator.

The Sun's Declination varies as the year progresses, reaching +23.5° around June 21, and −23.5° around December 21. In the northern hemisphere, these are the dates of the summer and winter solstices respectively; winter and summer are reversed in the southern hemisphere. At the summer solstice, from a given latitude, the Sun stands at its highest for the year above the horizon at noon, while in winter it will be at its lowest.

FIRST POINT OF ARIES

The Sun's apparent motion along the ecliptic is used to define the prime meridian of Right Ascension. This is the meridian drawn through the point where the Sun's centre crosses the celestial equator as it moves northward. This point, known as the First Point of Aries (from its position in ancient Greek times) moves very slightly westward each year as a result of precession, the 'wobbling' of the Earth's axis. Thus the First Point of Aries, at RA 0h 00m 00s Dec 0° 00m 00s, currently lies in Pisces.

The points at which the Sun crosses the celestial equator also define the equinoxes: the northern hemisphere vernal (spring) equinox is around March 21 as the Sun heads northward in our skies, while the autumnal equinox is the point where the Sun crosses the celestial equator as it heads south around September 23. The seasons are, of course, reversed if viewed from the southern hemisphere.

ALTITUDE AND AZIMUTH

The position of any celestial body at a given moment can be conveniently described in degrees, minutes and seconds of altitude above the horizon and azimuth from due north. These altazimuth coordinates continually change as Earth rotates, and objects rise or set.

Altitude can range from 0° on the horizon, to 90° at the zenith, directly overhead. Azimuth is measured from 0° due north, increasing through 90° due east, 180° due south and so on.

ALTITUDE AND AZIMUTH
Horizontal coordinate system based on an observer's horizon and north point. Altitude is measured north or south of the observer's horizon. Azimuth is measured eastward of due north. The point directly above the observer's head is called the zenith.

◄ 40–41 Phases of the Moon

◄ 155 RA and Dec

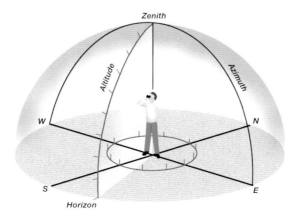

PRECESSION

At the moment, the Earth's axis of rotation points directly to the north pole star, Polaris, in the northern skies. However, over many years, the direction the axis points to will change as the Earth's axis slowly precesses or wobbles.

Over 25,800 years, the Earth's axis will describe a complete circle causing the position of the 'fixed stars' to move.

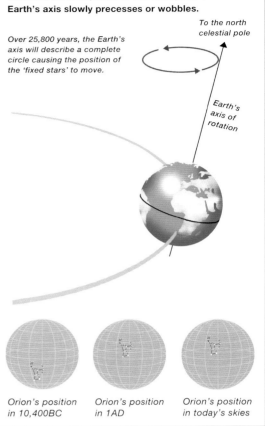

Orion's position in 10,400BC   *Orion's position in 1AD*   *Orion's position in today's skies*

Objects rise in the east, reaching their maximum altitude due south in the northern hemisphere and due north in the southern. The great circle passing through an observer's zenith and crossing the observer's horizon due north and due south is the celestial meridian, and objects attain their greatest altitude as they cross this meridian.

PRECESSION

Subtler changes in the sky's appearance occur over yet longer timescales. A small nutation – 'nodding' back and forth – of Earth's axis occurs in response to the Moon's gravitational tug over a period of 18.6 years.

Polaris has not always been the star closest to the north celestial pole. Around the time of the Pharaohs, alpha Draconis (Thuban) was the closest, reasonably bright, star. Deneb and Vega can also, in their turn, be north pole stars over the course of millennia. This is because the pulls of both Sun and Moon combine to produce the large-scale motion of precession. Over the course of 25,800 years the direction in which the Earth's axis points, while remaining tilted at 23.5° to the orbital plane, turns a complete circle around the heavens.

Movements due to precession require that star positions in the standard RA and Dec grid be updated regularly. Astronomers make these updates at 50-year intervals, and the next is due to take place in the year 2000.

PLANETARY MOVEMENTS

The five major planets that were known to ancient astronomers because they are visible with the naked eye – Mercury, Venus, Mars, Jupiter and Saturn – move against the fixed star background; indeed, the word planet derives from the Greek word 'planetos' meaning 'wanderer'. The apparent motions are determined by the planets' positions relative to Earth's orbit. All the major planets have orbits that lie close to the ecliptic plane.

INFERIOR PLANETS

Mercury and Venus have orbits around the Sun that lie inward from that of Earth, and are described as being inferior planets. Both appear to move back and forth along lines centred on the Sun, reaching their most prominent showing at greatest elongation east or west of the Sun because they are out of the solar glare. Eastern elongations occur in the evening sky, western elongations in the morning. Mercury never reaches an elongation of more than 27° (and Venus of 47°) from the Sun. Mercury is very difficult to observe as it is never visible more than two hours after sunset or before sunrise.

During their orbits, Mercury and Venus pass through what astronomers call inferior conjunction when they lie between the Sun and Earth and through superior conjunction when they lie beyond the Sun. At rare intervals, when the alignment at inferior conjunction is just right, you can see Mercury or Venus as black spots passing across the Sun. This is called a transit.

As a result of their changing angles in relation to the Sun, and to the moving Earth, Mercury and Venus show different phases, like the Moon does, when observed through a telescope. Their shape ranges from full, around superior conjunction, through half-phase at elongation and crescent around inferior conjunction.

SUPERIOR PLANETS

Mars, Jupiter and Saturn as superior planets have orbits that lie outside that of Earth. Superior planets emerge from conjunction behind the Sun into the dawn sky at the beginning of their apparitions, rising gradually earlier each night. For several weeks, they appear to move eastward relative to the star background, this apparent direct motion gradually slowing as the Earth, on its faster 'inside track', catches up with the superior planets. As the Earth and the superior planets come into line, direct forward motion seems to stop altogether and then appears to reverse westward for a short period of backward motion. This is called retrograde motion.

The point at which we catch up so that the superior planet is exactly 180° from the Sun in Earth's sky is termed opposition. At opposition, we are closest to the planet, and its disk diameter and magnitude will be at their greatest. Planets at opposition rise at sunset, reach the meridian at midnight and set at sunrise.

After opposition, Earth overtakes the planet, and the backward motion of the planet with respect to the Earth slows, then direct motion resumes as we pull away. The apparition ends as the planet becomes lost in the evening twilight glare, eventually reaching conjunction beyond the Sun again.

Mars has an orbital period of 687 days, returning to opposition at intervals of 26 months, Jupiter a 12-year orbit with successive oppositions 13 months apart, and Saturn a sluggish 29-year orbit, coming to opposition every 12.5 months.

HISTORY OF NAVIGATION
A major impetus behind the accurate mapping and cataloguing of the stars in the early days of astronomy was their use in navigation. Viking, Polynesian and other early seafarers are believed to have used the altitude of the pole star relative to the rigging on their boats as a means of determining latitude.

Determination of longitude was a more difficult problem, eventually solved in the 18th century by the invention of reliable marine chronometers that, coupled with tables of stellar positions given in annual nautical almanacs, could be used to find a ship's geographical position.

The ancient Egyptians used the seasonal appearance of the sky to gauge their agricultural calendar. The annual Nile floods, which deposited fresh, fertile soil for planting, coincided with the time when the Pleiades made their first appearance in the pre-dawn: this heliacal rising, recorded by astronomers, was an important event.

Even though they are not close to each other physically, planets can appear close when seen from the Earth. This photograph is of an early evening conjunction between Mars, Saturn and Jupiter.

Observing the planets 160 ▶

Orion 168–169 ▶

DIMINISHING DARKNESS

Modern telescopes are sensitive enough to see objects as faint as a candle flame from a distance of 8km, but to see objects thousands or even millions of light years away astronomers need to look through clear, dark skies. Astronomers worldwide are fighting to preserve something that belongs to us all – the night sky.

BY PAM SPENCE

Above: This satellite image of the world at night shows white city lights; yellow natural gas flares associated with oil fields; and red slash and burn agricultural fires (such as those in the tropical forest region of equatorial Africa).

" Looking up at the night sky, you should be able to see as many as 3000 stars. The Milky Way – the glow of hundreds of thousands of stars in the plane of our galaxy – ought to stretch right across the sky. If the constellation of Andromeda is above your horizon, you should be able to see the fuzzy blur of the Andromeda galaxy, 2,250,000 light years away and the farthest object visible to the naked eye. From farther south, you should be able to see the Orion nebula quite easily without using a telescope or binoculars. The chances are, however, that you will see none of these things because of the effects of light pollution – the unwanted spill of light from street lights, security lights, floodlights, advertising signs and everything else for which we use lights.

Any light that shines up into the sky reflects off water droplets and dust particles, so causing 'skyglow', the glow that hangs over cities and towns and that can affect the colour of the sky and how much you can see for many kilometres. Astronomers have estimated that a town with a population of about 3000 will spill enough light into the sky to be seen 10km away, while a city with a population of one million may cause problems for people looking at the night sky over 100km away.

Most street lights work by passing an electric current through a gas so that it glows, and different gases glow in different colours. Each type of lighting causes different problems for astronomers and the general public. Low-pressure sodium fittings – large, orange lights that spill their glow in all directions – interfere with both photographic and visual astronomy, but as they emit light in only a very narrow band, it is possible to counter this effect on astronomical spectra with filters or by computer processing. Unfortunately, the lights' yellow glow makes everything look the same colour, so it is difficult, for instance, to pick out a red car from those of other colours in a car park.

The wonderful sight of the stars is part of our heritage: I would not like to have to explain to my grandchildren what a star in the night sky looked like.

High-pressure sodium fittings have been developed that have less effect on how we see colours, but since they emit light at a wider range of wavelengths, it is more difficult for astronomers to subtract from observations.

Astronomers are not asking everyone else to live in darkness – we need to get home safely, too, but light pollution is unnecessary and wasteful. Street lighting should be just for lighting roads; cars are not driven 100m above a road. Fittings that throw a large proportion of their light up into the sky waste energy and cost money. If a 100-watt bulb in your home throws nearly half of its light where you don't want it, why not fit a better shade and use a 60-watt bulb instead? The same applies to street lights.

Some newer lights have greatly improved fittings that have been designed to minimize light spill into the sky by reflecting light downward where it is needed. The darkness above these full cut-off fittings can be dramatic. So why don't all lighting authorities fit downward-shining lights? The original excuse was that the fittings were more expensive, but the costs are now rapidly coming down. Some lighting authorities, however, still resist change: after all, astronomers are a minority and 'you are the only person to complain about this light' is a common answer to their requests for change. But the night sky belongs to everyone – astronomers are just more aware of the situation than most people.

All over the world, astronomical observatories have been moved away from city centres, only to have urban sprawl and its associated light creep up to their walls. The Royal Greenwich Observatory, founded in London by Charles II in 1675, was moved into the countryside near the coast to escape the polluted London sky in the 1940s. By the late 1960s the new site was already suffering from light pollution from the nearby towns and villages so British astronomers built telescopes on

La Palma, in the Canary Islands, and Hawaii, but both sites are now threatened. In California, the telescopes on Mount Wilson are completely swamped by Los Angeles' lights. Only when the city was under total black-out during World War II could the American astronomer Walter Baade do his amazing work there on resolving individual stars in the Andromeda galaxy.

The Japanese national observatory, built on the outskirts of Tokyo in 1878, moved away from the city in the 1920s. Today the sky is so bright near the observatory that little research is possible. The Japanese are building a telescope on Mauna Kea, Hawaii, as are astronomers from many countries, but Hawaii's sky will get lighter and lighter unless action is taken to stop the spreading skyglow. Soon there will be nowhere else on Earth for astronomers to go.

In 1987, the Star-Watch programme was started in Japan, with the aim of monitoring night-sky brightness and air pollution and to increase public awareness of the problems. Star-Watch has since spread to the US and the UK. Generally, people are asked to look at the Pleiades, mark the stars they see on a map and send it to the programme.

The International Dark-Sky Association has members in over 65 countries, and astronomers are campaigning for better lighting. In the US, Canada, Japan, Germany, France, Italy and New Zealand campaigns have resulted in laws to control lighting at county, town, city or state level. In 1997, the Italian government declared 4 October National Light Pollution Awareness Day.

Helping to reduce skyglow can be as simple as turning off a light, changing a bulb for one of a lower wattage, or pointing a fitting down instead of up. As towns expand, more and more street lights appear, and the worse the problem gets. Unless we make lighting authorities aware of the long-term cost benefits of efficient light fittings, they will continue erecting inefficient, wasteful ones, and the skyglow will get brighter and brighter, until no-one can see the stars at all. The sight of the stars is part of our heritage: I would not like to have to explain to my grandchildren what a star in the night sky looked like. **"**

The city lights and rural fires show our presence on the planet in this mosaic of images from the US Air Force weather satellites. (The fuzzy light at the top of the image shows the northern lights.) Most of the light leakage into space comes from street and building lights in Europe, North America and Eastern Asia. This image clearly illustrates the constant battle with light pollution – millions of people have no dark sky and are denied the thrilling sights of a meteor shooting across the sky and the resplendence of the Milky Way. We have wrapped the Earth in a glowing fog.

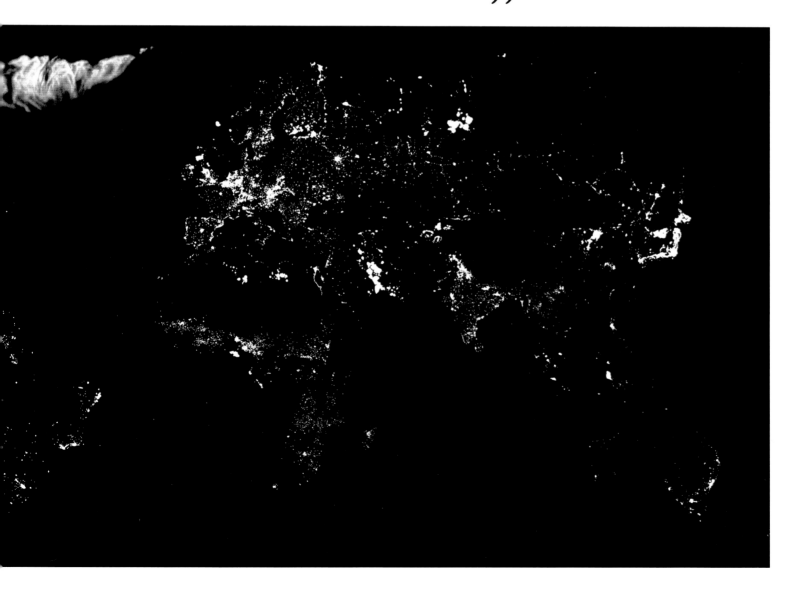

SKYWATCHING

THE NAKED EYE, A STAR MAP AND A TORCH ARE ALL
YOU NEED TO START SKYWATCHING, BUT BINOCULARS
AND A TELESCOPE WILL REVEAL A LOT MORE DETAIL.

When you are observing the Moon, planets and stars, much depends on the stability of the atmosphere. On transparent, frosty nights when the stars are twinkling violently, you will find that the atmosphere is unfavourable for resolving fine details. Paradoxically, the best seeing for such observing often occurs on calm, hazy nights. Sometimes, it may help to use averted vision – directing the eye slightly to one side of the object – to bring out detail at the eyepiece: the retina's periphery is more sensitive to faint light than its centre.

OBSERVING THE MOON
The Moon is our closest astronomical neighbour and much can be revealed by observing it with just the naked eye. You can see the maria, lava plains (below), which make up the patterns of dark patches, familiar to many people as the face of the 'Man in the Moon'. You can also see the large bright rays that radiate out of certain craters, such as Tycho and Copernicus. The best detail can be seen at the terminator, the line between light and dark (below), where the Sun's rays pick out the lunar mountains and plunge the maria into shadow.

The most striking feature of the Moon visible with the naked eye is its changing phases. It takes just under 30 days for the Moon to change from new moon to full moon. Following the Moon's progress during the month, and in particular observing how the features change along the terminator, is a fascinating pastime.

With binoculars, you can see details of individual craters and mountains and the shadows that they cast. Still more detail is revealed with a 100mm reflector telescope. You can see the smallest of craters and make out the details inside them, as well as the ejecta lying around the outside.

OBSERVING THE PLANETS
The planets that are visible with the naked eye are those known about since antiquity: Mercury, Venus, Mars, Jupiter and Saturn.

Mercury never appears far from the Sun in the sky. For northern hemisphere observers, spring evening and autumn morning skies offer the best chance to catch the planet at magnitude zero in the twilight, and binoculars will certainly help, although care must be taken not to catch the Sun's rays through them.

You need binoculars in order to make out details of the planets. They enable you to see the different phases of Venus, just as you can with the Moon. You can also see the moons of Jupiter that were discovered by Galileo in the 17th century and the striking ring system around Saturn.

However, serious planet watching requires a telescope. Distant Uranus and Neptune, at respective magnitudes +6 and +8, require telescopes to be easily detected. The most rewarding planet for amateur observers with telescopes is giant Jupiter, with its continually changing cloud features, seen on a flattened disk with a large diameter of about 45 arc seconds. Dark belts and

The north-east quadrant of the Moon, showing lava plains. Mare Imbrium, the 'Sea of Showers', lies across the terminator (the line between light and dark) and Mare Serenitatis, the 'Sea of Serenity', is on the lower right.

lighter zones are often interrupted by storms or spots. The Great Red Spot has been followed in amateur-sized telescopes for centuries. Much can be revealed about Jovian weather systems by making series of disk drawings and timing the passage of features across the planet's north–south meridian.

Saturn has blander cloud features, but rare white spot outbreaks, as in 1990, can be interesting. The ring system, when favourably presented, is one of the outstanding sights in a moderate-sized amateur telescope.

The phases of the brilliant planet Venus (usually prominent as a morning or evening 'star') can be seen very clearly through a small telescope, ranging from full on the far side of the Sun, to half-phase around elongation and a crescent when more or less between the Sun and Earth.

Mars usually requires a large telescope to reveal its dark markings and bright polar cap. These change with the Martian seasons, and unpredictable dust storms or clouds may obscure the more permanent features. Seasonal changes occur over a period of weeks, and some features also change from one apparition to the next.

OBSERVING COMETS

Comets that are bright enough to be seen with the naked eye are very rare. Hyakutake and Hale-Bopp are spectacular recent exceptions.

However, observers with binoculars or telescopes, who are skilful in the use of sky charts, can see up to 20 each year. In a small telescope or binoculars, the most obvious feature of a faint comet is the bluish circular or teardrop-shaped coma surrounding the nucleus.

Halley's comet, seen here in 1986, is a glorious example of a bright comet visible to the naked eye. The nucleus is hidden by the coma, while its gas or ion tail streams outward away from the Sun.

Equipment 164–165 ▶

Stellar magnitudes 166–167 ▶

Star and Moon maps 170–183 ▶

Sometimes the coma is very diffuse, meaning that the comet is sometimes difficult to make out against the sky background. Brighter comets may have a tail or tails. The gas tail is usually quite straight and glows bluish or white. The yellowish dust tail, if present, may appear curved.

OBSERVING METEORS

One of the simplest and most rewarding fields of naked-eye amateur astronomy is meteor observing. When the major dependable annual showers (for example the Perseids on August 12 or the Geminids on December 13–14) are at their most active, watches several hours long can be very productive.

Ideally, you should be sat comfortably in a deck chair, wrapped against the cold, and looking into the sky to one side of the radiant, the point in the sky from which shower meteors seem to emerge. During the watch, note the appearance times and magnitudes for any meteors. While many observers work alone, group meteor watching is a popular club activity.

OBSERVING AURORAE

One of nature's most awesome spectacles, best seen with the naked eye, is the aurora, a display of rippling bands of colour towards the poles of the Earth. This is caused by charged particles from the Sun cascading down the Earth's magnetic field lines and interacting with the atmosphere.

Displays can be somewhat unpredictable but, following violent solar activity, the aurorae may reach to lower-than-normal latitudes.

OBSERVING THE SUN

Never look directly at the Sun with the naked eye or directly through any optical instrument: permanent eye damage is the only likely outcome. Sensible precautions can be taken that will allow you to observe the Sun with the smallest of telescopes but do not be tempted to use the inadequate filters that are often provided with beginners' telescopes.

The safest method of observing the Sun is by projection. You will often be able to see sunspots, frequently in groups, on the projected disk as dark areas mottling the bright background. Large spots may have a central dark umbra, surrounded by a lighter region of penumbra. The Sun's appearance changes from day to day as it rotates and groups of spots break out or decay.

You can make daily sketches of the changing appearances and positions of spot groups and build sunspot counts into a record showing the variations of solar activity over periods of months and years.

OBSERVING STARS

Observers of variable stars use the eye's ability to gauge differences in brightness. There are about 50 naked-eye stars which vary markedly, such as delta Cephei. There are hundreds more variable stars that can be followed with binoculars, such as Mira in the constellation Cetus.

However, you need a telescope for any detailed study of the stars. Telescopic variable stars number in the thousands. Many of the stars in the sky, when viewed in a telescope, appear to be double or multiple systems. Some pairings reveal beautiful colour contrasts. Albireo in Cygnus, for example, couples a magnitude +3 orange star with a bluish magnitude +5 secondary. The two lie 34 arc seconds apart, and can be separated using small telescopes at x30 magnification.

There are dozens of line-of-sight doubles in the sky, which look as if they are close together but are physically very distant from each other. Physically associated pairs, true binary stars, are also fairly common. Among the finest

A Perseid meteor is shown streaking across the sky. The Perseid meteor shower is seen each year on about August 12, and seems to originate in the constellation Perseus.

The aurora is a magnificent sight with rippling bands of green and red light moving across the night sky. Aurorae are seen in both the northern and southern hemispheres, and their strength and frequency are related to the sunspot cycle.

is Castor (alpha Geminorum), whose two major components (the system contains six stars in all), of magnitudes +2 and +3, lie less than four arc seconds apart. Epsilon Lyrae comprises two binaries, four arc minutes apart, each separated by less than 2.5 arc seconds.

DEEP SKY OBSERVING

Some of the more prominent objects in deep space, such as the Pleiades star cluster or the Andromeda galaxy (M31), can be seen with the naked eye. A pair of 10x50 binoculars will enable you to see objects down to around magnitude +8 and will resolve some open clusters into individual stars.

However, deep sky observing really requires a telescope. The more powerful the telescope, the deeper you will be able to see. Thousands of open and globular star clusters, diffuse and planetary nebulae and galaxies are there to be observed.

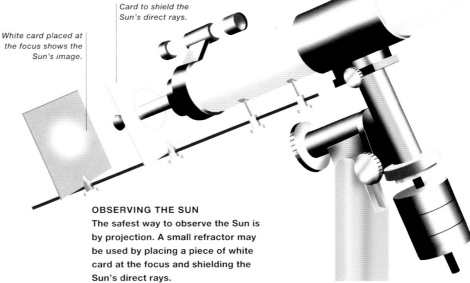

Card to shield the Sun's direct rays.

White card placed at the focus shows the Sun's image.

OBSERVING THE SUN
The safest way to observe the Sun is by projection. A small refractor may be used by placing a piece of white card at the focus and shielding the Sun's direct rays.

EQUIPMENT

THERE IS A WEALTH OF EQUIPMENT AVAILABLE TODAY FOR AMATEUR ASTRONOMERS, FROM A BASIC PAIR OF BINOCULARS TO HI-TECH COMPUTER PROGRAMS.

With the equipment available today, it is sometimes amateur astronomers who are the first to spot the events occurring in the heavens.

BINOCULARS

A great range of interesting objects lie within reach of binoculars. On a cost basis alone, a good pair of binoculars is a better investment than a poor-quality telescope.

Binoculars are described in terms of their magnification and aperture. Thus, a pair described as '7x35' has a magnification of x7, and lenses of 35mm aperture.

A rule that applies equally to binoculars and telescopes is that the larger the aperture, the greater the light-gathering power and, hence, ability to detect faint objects.

The ideal beginners' binoculars are 7x35 or 10x50, which are generally reasonably cheap, and can be hand-held without too many problems. Many photographic suppliers sell simple adapters that allow binoculars to be clamped to lightweight camera tripods.

Many more advanced observers often prefer to use large aperture binoculars such as 15x80, for comet sweeping or for detecting faint variable stars. These are both expensive and cumbersome, and they need a solid tripod to support them.

A homemade Dobsonian mount telescope can reveal many delights of the night sky.

TELESCOPES

Just as binoculars reveal sights not visible to the naked eye, so telescopes, with their greater magnification and light-gathering power, will open up the universe even more. Telescopes work by having a lens or mirror to gather as much light as possible and bend it to a focus in an eyepiece. They are described in terms of the diameter of the lens or mirror, such as 100mm. There are three types of telescope commonly used by amateur observers: refractors, reflectors and catadioptrics.

Refractors employ a glass objective lens to gather light. This type is best for resolving detail in near space, such as the planets or the Moon. An aperture of at least 75mm is required for anything more than observing the craters on the Moon, Saturn's rings or the Galilean satellites. Reflectors use mirrors to collect light and focus it at the eyepiece. The Newtonian reflector has long been a popular amateur instrument. Catadioptric telescopes offer a compact design in which the light path is folded into a short tube by using both lenses and mirrors.

TYPES OF TELESCOPE
There are two basic types of telescope: the refractor, which uses lenses to focus the light from far away objects and the reflector, which uses a mirror.

REFRACTOR

Incident light ray

Objective lens

Telescope tube

Refracted light ray

Eyepiece lens

REFLECTOR

Incident light ray

Eyepiece lens

Telescope tube

Primary mirror

Secondary mirror

Reflected light ray

MOUNTINGS

No matter how good your telescope is optically, it will be limited by a poor-quality mount. Particularly for large aperture instruments, a sturdy mount is essential. There are two types of mount: altazimuth and equatorial.

Altazimuth mounts allow the telescope to be slewed up and down and side to side in any direction. They are commonly used for small, portable instruments and require the minimum of setting up. In recent years, a variation, the Dobsonian mount, has become popular, offering a cheap and portable means of mounting large aperture reflectors.

The problem with altazimuth mounts is that the target will stay in the field of view for only a very short time before

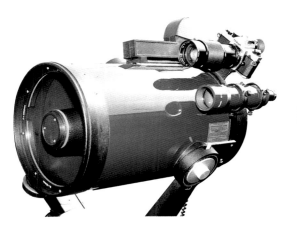

Far left: A camera can be fixed 'piggy back' to a telescope so that the drive system can move the camera to follow the stars.

Left: A camera can be fixed to the telescope at prime focus: essentially using the telescope as a large camera lens.

being carried away by the Earth's rotation. This is a particular nuisance when working at high magnifications. The solution is to have a mount that has one axis carefully aligned to the pole, allowing the telescope to be driven westward at the same rate as the apparent drift of the stars (the sidereal rate). Such mounts are called equatorial.

The axes of equatorial mounts are often fitted with setting circles, marked out in degrees, to help in locating objects. These allow you to enter the RA and Dec of a particular star and it will focus on the right part of the sky.

FINER POINTS

A telescope's resolving power depends chiefly on its aperture: for example, a 75mm refractor should be able to split stars separated by 1.5 arc seconds, while a 150mm reflector will resolve objects 0.8 arc seconds apart (equivalent, under good conditions, to lunar features around 2km in diameter).

ACCESSORIES

There are a vast array of accessories available for the observer. Some are more useful than others. Among the most important is, of course, a torch. A torch dimmed with a red cover has least effect on your night vision.

When hunting for faint objects, it is useful to have star charts showing stars down to at least the binocular magnitude limit. Using these, you can also find objects by star-hopping: using a bright star as a starting point, then following patterns of fainter stars to reach the target.

ASTROPHOTOGRAPHY

For astrophotography, a single lens reflex (SLR) camera is needed for taking time exposures, preferably by mechanical rather than battery-powered means. The standard 50mm lenses provided with most SLR cameras are useful for wide field photography, as are those of 28mm focal length.

In order to keep focused on the constellation you are trying to photograph, you must adjust the camera to counter the rotation of the Earth. You can either attach the camera to the outside of telescope tubes (above left), allowing you to use the telescope's drive to follow the stars. Small, independent camera drive systems are also available. Alternatively, you can take photographs through the telescope (above right) – essentially using the telescope as a very large, interchangeable lens – special adapters fitting into the eyepiece mount are available.

CCDS

A revolution in amateur astronomical imaging has been brought about by the increasing availability of charge coupled devices (CCDs). For many observers, the CCD chip, an array (lattice-shaped group) of tiny, very sensitive photomultipliers, has replaced film as the ideal medium for permanent recording.

COMPUTER PROGRAMS

Much of the 'CCD revolution' in amateur astro-imaging has been brought about by parallel advances in personal computers. Many observers process their CCD images using standard commercial computer programs.

Astronomical software abounds, too. There are a great many 'planetarium' programs available that show the appearance of the sky down to selectable magnitude limits. Some are coupled to wider databases of astronomical information and images, and can be useful learning tools, providing plenty to explore on a cloudy night.

Computer programs can help in planning a night's observing. Some programs can even automatically guide your telescope for you. Computerized RA and Dec axes on some telescope mounts with motor drives allow automatic object finding.

The Internet can provide a wealth of further data. Many astronomical institutions and clubs are now 'on-line', and the information resources that are available via modem to a home computer are vast.

HI-TECH OBSERVING

While a supernova has not been seen in our own galaxy for some 300 years, many observers survey other galaxies for occasional stellar cataclysms. Amateurs are often the first to spot the latest developments. Some automatic amateur supernova searches now exploit computerized telescope-pointing, with CCD imaging capturing faint magnitudes. Image analysis can be done, within minutes, in warmth and comfort indoors.

Considerable investment in both time and resources is entailed, but the scientific rewards that are gained are great: early alerts about new supernovae allow professional astronomers to obtain important information about how massive stars end their existence.

Catadioptric telescopes combine mirrors and lenses and are compact and portable instruments.

CONSTELLATIONS

OVER THE CENTURIES, PEOPLE HAVE BEEN FASCINATED
BY THE PATTERNS OF STARS IN THE SKY AND HAVE
INTERPRETED THEM IN A NUMBER OF WAYS.

*Many old star atlases are
beautiful works of art depicting
the mythological figures of
the constellations.*

In 1922 88 'official' constellations were agreed upon internationally, and in 1930 the International Astronomical Union adopted boundaries for these constellations, designed by the Belgian astronomer Delporte.

CATALOGUING THE STARS

Many stars have common names: some of these are Greek or Roman in origin, but many more are Arabic. On our star maps the names of all stars brighter than magnitude +1.5 are shown, as well as those of a few well-known fainter stars like Algol, Mira and Denebola.

Stars also have other designations – Greek or, sometimes, Roman letters and numbers. The first system used was introduced by Johann Bayer in his atlas *Uranometria*, published in 1603. In general, Bayer assigned the first letter of the Greek alphabet, alpha (α), to the brightest star in a constellation, beta (β) to the second brightest, gamma (γ) to the next, and so on. But there are examples where Bayer was not very strict with this 'rule', such as Sagittarius where α and β are dim stars in the southern part of the constellation.

The Greek alphabet has only 24 letters – not always enough for all of the stars in a constellation – and Bayer sometimes used one letter for a group of stars, and numbered the individual stars within the group as well, so, the stars in Orion's shield are shown as π^1, π^2, π^3, etc.

The Greek letter is usually followed by the genitive form of the Latin constellation name or its three-letter abbreviation. Thus Rigel, in Orion, is also called β Orionis or β Ori.

Star numbers are usually known as 'Flamsteed numbers', from a 1780 French catalogue by the English astronomer, John Flamsteed. Many stars catalogued by Bayer also have 'Flamsteed' numbers, thus Rigel is also 19 Orionis or 19 Ori.

In later periods, the naming of stars was expanded to include fainter stars by several other astronomers using Roman letters, both capitals and lower case.

Variable stars have their own system of numbers. In the 19th century the German astronomer, Friedrich Argelander, used the capital letters R to Z, followed by the constellation name, for variables. When he ran out of these letters, he continued with double capitals. This results in a series of 334 letters or double letters. After that, the list continues with V335, V336, etc.

There are various ways of interpreting the patterns of stars and of cataloguing the individual stars. There is also a method of denoting their brightness.

PLOTTING THE CONSTELLATIONS

Many of the names of modern constellations like Orion, Cepheus, Hercules and Andromeda originate from Greek mythology. All these mythological characters have their own tale, sometimes in several different versions and often connected with the tales of other constellations.

A part of the southern sky was unknown to the Greeks, Babylonians and Egyptians and so stayed devoid of constellations until the 16th century when seafarers started to explore the southern oceans. In the north, star-map makers often made up their 'own' constellations in the empty spaces between the classical groups.

In the middle of the 18th century, a French astronomer, De Lacaille, added a series of 'modern' constellations to the southern sky, and on a star map today you will still find Orion and Hercules, but also a pendulum clock, a telescope and a microscope.

STELLAR MAGNITUDES

Even a casual look into a clear night sky will show that the stars come in a range of brightnesses. In ancient times, astronomers ranked them in terms of magnitude, the brightest stars being of the first magnitude, those slightly less bright of second magnitude, and so on down to the naked-eye limit of magnitude +6. The fainter the star, the larger the magnitude.

Later, a more precise definition was conceived: each ascending magnitude step is 2.512 times fainter. A difference of five magnitudes corresponds to a brightness factor of 100.

Some of the more brilliant objects have magnitudes (commonly abbreviated to mag.) described by a negative prefix. Sirius, the brightest star, is of mag. –1.5. The full moon has a magnitude of the order of –12, while the Sun is a dazzling mag. –23. Venus is the brightest of the planets around mag. –4, Jupiter mag. –2.

THE 88 CONSTELLATIONS

Latin name	Abbr.	English name	Latin name	Abbr.	English name
ANDROMEDA	And	Andromeda	LACERTA	Lac	Lizard
ANTLIA	Ant	Air Pump	LEO	Leo	Lion
APUS	Aps	Bird of Paradise	LEO MINOR	LMi	Small (Lesser) Lion
AQUARIUS	Aqr	Water Carrier	LEPUS	Lep	Hare
AQUILA	Aql	Eagle	LIBRA	Lib	Scales
ARA	Ara	Altar	LUPUS	Lup	Wolf
ARIES	Ari	Ram	LYNX	Lyn	Lynx (Bobcat)
AURIGA	Aur	Charioteer	LYRA	Lyr	Lyre
BOÖTES	Boo	Herdsman	MENSA	Men	Table Mountain
CAELUM	Cae	Graving Tool	MICROSCOPIUM	Mic	Microscope
CAMELOPARDALIS	Cam	Giraffe	MONOCEROS	Mon	Unicorn
CANCER	Cnc	Crab	MUSCA	Mus	Fly
CANES VENATICI	CVn	Hunting Dogs	NORMA	Nor	Level (Carpenter's Square)
CANIS MAJOR	CMa	Large (Greater) Dog	OCTANS	Oct	Octant
CANIS MINOR	CMi	Small (Lesser) Dog	OPHIUCHUS	Oph	Serpent Holder (Serpent Bearer)
CAPRICORNUS	Cap	Sea Goat	ORION	Ori	Orion
CARINA	Car	Ship's Keel	PAVO	Pav	Peacock
CASSIOPEIA	Cas	Cassiopeia	PEGASUS	Peg	Pegasus
CENTAURUS	Cen	Centaur	PERSEUS	Per	Perseus
CEPHEUS	Cep	Cepheus	PHOENIX	Phe	Phoenix
CETUS	Cet	Whale (Sea Monster)	PICTOR	Pic	Painter's Easel
CHAMAELEON	Cha	Chameleon	PISCES	Psc	Fishes
CIRCINUS	Cir	(Drawing) Compasses	PISCIS AUSTRINUS	PsA	Southern Fish
COLUMBA	Col	Dove	PUPPIS	Pup	Ship's Stern
COMA BERENICES	Com	Berenice's Hair	PYXIS	Pyx	(Mariner's) Compass
CORONA AUSTRALIS	CrA	Southern Crown	RETICULUM	Ret	Retile (Net)
CORONA BOREALIS	CrB	Northern Crown	SAGITTA	Sge	Arrow
CORVUS	Crv	Crow	SAGITTARIUS	Sgr	Archer
CRATER	Crt	Cup	SCORPIUS	Sco	Scorpion
CRUX	Cru	Southern Cross	SCULPTOR	Scl	Sculptor
CYGNUS	Cyg	Swan	SCUTUM	Sct	Shield
DELPHINUS	Del	Dolphin	SERPENS CAPUT/CAUDA	Ser	Serpent (Head/Tail)
DORADO	Dor	Swordfish (Goldfish)	SEXTANS	Sex	Sextant
DRACO	Dra	Dragon	TAURUS	Tau	Bull
EQUULEUS	Equ	Colt	TELESCOPIUM	Tel	Telescope
ERIDANUS	Eri	(River) Eridanus	TRIANGULUM	Tri	Triangle
FORNAX	For	Furnace	TRIANGULUM AUSTRALE	TrA	Southern Triangle
GEMINI	Gem	Twins	TUCANA	Tuc	Toucan
GRUS	Gru	Crane	URSA MAJOR	UMa	Great Bear
HERCULES	Her	Hercules	URSA MINOR	UMi	Small (Lesser) Bear
HOROLOGIUM	Hor	Pendulum Clock	VELA	Vel	(Ship's) Sails
HYDRA	Hya	(Female) Water Snake	VIRGO	Vir	Virgin
HYDRUS	Hyi	Lesser (Male) Water Snake	VOLANS	Vol	Flying Fish
INDUS	Ind	(American) Indian	VULPECULA	Vul	Fox

Differences in magnitude result from the intrinsic brightness of stars, and their varying distance. The apparent magnitude of a star – as we see it in our skies – does not necessarily indicate its absolute magnitude. A very luminous distant star may appear less brilliant than a nearby dimmer one.

A star's distance gives the absolute magnitude: the apparent brightness at a distance of 32.6 light years (10 parsecs). For example, Rigel has absolute magnitude –7 and apparent magnitude 0; our Sun has absolute magnitude of +5.

α	alpha	ι	iota	ρ	rho
β	beta	κ	kappa	σ	sigma
γ	gamma	λ	lambda	τ	tau
δ	delta	μ	mu	υ	upsilon
ε	epsilon	ν	nu	φ	phi
ζ	zeta	ξ	xi	χ	chi
η	eta	o	omicron	ψ	psi
ϑ	theta	π	pi	ω	omega

Orion 168–169 ▶

Star and Moon maps 170–183 ▶

ORION

ORION IS ONE OF THE MOST FAMOUS AND BEAUTIFUL
CONSTELLATIONS AND ONE OF THE FEW THAT IS
VISIBLE FROM ALL OVER THE WORLD.

The central part of the constellation of Orion showing the
positions of the details opposite. 1 is M42, the Orion nebula;
2 is M43 another area of bright nebulosity; 3, near Orion's
belt, is the Horsehead nebula and 4 is the red supergiant
star Betelgeuse.

*In dark skies and with long
exposures, photographs can pick
out detail including the red colour
of Betelgeuse and the Orion
nebula in the 'sword'.*

Orion the hunter is distinctive in the sky because of the
three bright stars of his belt, suspending his 'sword' in
which lies the famous Orion nebula.

THE MYTH OF ORION

In Greek mythology, Orion was the giant son of the sea god
Poseidon and the nymph Euryale. He was a great hunter
but felt no compassion for the animals he was hunting. He
boasted that he was the greatest hunter on Earth and would
be able to kill all the animals. When the Earth goddess Gaea
heard this she sent a scorpion to sting Orion. In some stories
he died when he was stung, but in other versions he was
rescued by Ophiuchus, the serpent holder, who gave him
an antidote and trampled the scorpion. In the sky, Orion is
fighting the Bull (Taurus) and is followed by the Greater Dog
and the Lesser Dog (Canis Major and Canis Minor).
Ophiuchus, the Serpent Holder, is seen with one of his feet
on Scorpius, the Scorpion.

WHERE CAN YOU SEE ORION?

The distinctive belt, marking the hunter's waist, sits close to
the celestial equator, which means that Orion is visible from
all parts of the inhabited world. To those in the northern
hemisphere, Orion lies near the southern horizon on winter
nights and for southern hemisphere observers, Orion is seen
near the northern horizon and is at his best in the summer.

Looking at Orion from Earth is looking outwards, away
from the galactic centre of the Milky Way to the next spiral
arm of the galaxy. It is notable as the single constellation
with most stars brighter than magnitude +2.

Orion contains much nebulous material from which stars
are currently forming. The region is strewn with objects of
interest, many of which can be seen with amateur equipment.

Betelgeuse (α Orionis) is the brightest red supergiant star
readily evident to the naked eye. The name comes from the
Arabic for 'Armpit of the Giant'. Betelgeuse lies at a distance
of 650 light years from Earth, and is around 20 times as
massive as the Sun. Its hydrogen-burning phase is over and
Betelgeuse is nearing the end of its life. It is expected to
explode as a supernova in a few million years.

While massive, Betelgeuse is also bloated – it is a true giant
with a diameter 1600 times that of our Sun. It is one of the
few stars that show a tiny disk to Earth-based telescopes;
use of interferometric techniques reveal the existence of hot
and cool areas on its surface. The average surface
temperature is around 3100K – cool by stellar standards.

Betelgeuse is a variable star, ranging from magnitude +0.4
to +1.2. The variations are irregular, but seem to occur about
every six years. The absolute magnitude of Betelgeuse is
around –6, its RA 05h 55.2m and Dec +07° 24.5m.

Rigel (β Orionis) is a blue supergiant and is the seventh
brightest star in the sky, easy to make out with the naked
eye. The name comes from the Arabic 'Left Leg of the
Giant'. It lies a distance of around 900 light years from
Earth and has a surface temperature of 12,000K. With a
mass 50 times that of our Sun, Rigel is expending its
nuclear fuel at a prodigious rate, and is expected to end its
life as a supernova in several million years, having first
passed through a phase as a red supergiant. The apparent
magnitude of this bright star is +0.1, its absolute
magnitude is –7.1 and its RA 05h 14.5m and Dec –08° 12m.

Orion Nebula (M42, NGC1976), lying within the 'sword', is
a bright (magnitude +5) emission nebula, some 1600 light
years away. Embedded within this region of dust and gas
are the hot young stars of the Trapezium. Ultraviolet
radiation from these stars ionizes the hydrogen gas and
radiation at specific wavelengths is given out as a result.
In photographs of the Orion nebula, reddish emission
at a wavelength of 656x10⁻⁹m (known as hydrogen-alpha)
is prominent; it often appears greenish to visual
observers because of the ionized oxygen emission. Its RA
is 05h 35m and Dec –05° 30m.

M42, part of the Orion nebula, is the most conspicuous
region of a vast complex of nebulosity that covers much of
Orion's sword. M42 is the bright central portion with a
diameter of about six light years. It is visible to the naked
eye on a clear, dark moonless night as a fuzzy 'star'. Seen
through binoculars it looks like an obvious hazy patch,
twice the apparent diameter of the Moon. The view through
a telescope reveals fine filaments of gas and much
structural detail. A dark nebula to the south, superimposed
on the bright emission of M42, is known as the 'Fish's
Mouth'. A detached portion of bright nebulosity, still further
south, is catalogued as M43 (NGC1982). Its RA is 05h 35m
and Dec –05° 30m.

The Horsehead Nebula and IC434 (B33, from a catalogue of dark nebulae produced by the American astronomer E E Barnard). The Horsehead is seen in strong contrast against the bright emission nebula IC434 near ζ Orionis at the eastern end of Orion's Belt. Although it shows up clearly in long-exposure photographs and CCD images, the Horsehead is elusive to the visual observer, even when very large telescopes are used. IC434 is thought to be part of the nebula complex 1600 light years away, while the Horsehead is the most prominent part of a dark nebula lying between it and our solar system at a distance of 1200 light years. The dark mass of the Horsehead is estimated to have a diameter of 11.4 light years. Its RA is 5h 41m and Dec –02° 35m.

M78 (NGC2067/2068) is a reflection nebula that lies some 1600 light years away. This is a challenging object to see through binoculars, but is well within the range of small telescopes. It is approximately two to three light years in diameter, illuminated by two stars of apparent magnitude +10, each a blue giant with absolute magnitude –1.5. Its RA is 05h 47m and Dec +00° 03m.

NGC2022 is a planetary nebula (one of only two in Orion), at a distance of 7200 light years. It is best seen through telescopes of 200mm aperture or greater and appears slightly oval, with a diameter of 28 arc seconds along its major axis. Like other planetary nebulae, NGC2022 with a magnitude +12.3 shines strongly in the greenish light of ionized oxygen at a wavelength of 500.7×10^{-9}m. The nebulosity is comprised of material shed from the outer atmosphere of the central star, during an earlier, red giant phase of its existence. Stripped of its outer layers, the central star has now become a white dwarf. Its RA is 05h 42m and Dec +09° 05m.

NGC2169 is an open cluster containing perhaps 20 stars in an area of about 8 arc minutes' diameter. This sparse grouping is best seen through a small telescope. Light from these stars, around magnitude +8, is dimmed somewhat by invisible dark matter lying between us and the cluster. Its RA is 06h 08.5m and Dec +13° 55m.

NGC1981 is a fine, scattered, open cluster containing about 20 stars, some as bright as magnitude +7 NGC1981 is best seen with a low power instrument – ideally binoculars – and lies just north of the Orion nebula (M42). Its RA is 05h 35m and Dec –04° 26m.

The Trapezium (ϑ¹ Orionis) is a multiple star system, whose members have formed recently (within the past 300,000 years) from the Orion nebula. The system takes its name from the four most prominent stars, grouped within 25 arc seconds of each other, and easy to make out through telescopes. Long-exposure photographs show the Trapezium stars to be the most prominent members of a small cluster. Ultraviolet radiation from these stars illuminates the central part of the Orion nebula, and is burning a hole through the dust rendering the stars visible from the Earth. Its RA is 05h 35.3m and Dec –05° 23m.

M43 is a bright area of gas and dust slightly separated from the main Orion nebula M42.

The Orion nebula, M42, is a region of dust and gas glowing because of the heat of newly born stars contained within it.

The Horsehead nebula is a cloud of dust and gas silhouetted by a brighter emission nebula behind. The Horsehead appears dark because there are no nearby stars to light it up.

Z **Orionis** is a binary star with components of magnitude +1.9 and +4.0 separated by a little more than two arc seconds. This pair can be resolved in small telescopes, and has an orbital period of 1500 years. Its RA is 05h 40.8m and Dec –01° 57m.

U Orionis is a Mira-type (long period) variable star, with an extreme range in brightness from magnitude +4.8 to +13.0, which varies at intervals of 368 days, quite typical for pulsating red giant stars of this type. Even at its brightest, when it often appears red, U Orionis can rarely be seen with the naked eye but can be seen through binoculars. Its RA is 05h 55.8m and Dec +20° 10m.

Betelgeuse is a red supergiant and the first star, apart from the Sun, which had its disk resolved.

Star and Moon maps 170–183 ▶

STAR AND MOON MAPS

STAR MAPS WILL HELP YOU TO FIND YOUR WAY
AROUND THE SKY TO ENABLE YOU TO DISCOVER A
MYRIAD FASCINATING OBJECTS.

When first looking up at the sky, look out for a bright star that will help you locate yourself on the map and then star-hop to more challenging constellations.

THE STAR MAPS

The maps on the following pages are specially designed for this book and show the whole sky, divided up into five double-page maps. Map 1 shows the northern part of the sky, with the celestial north pole in the centre, and Map 5 shows the south polar area. The other three maps depict the area around the celestial equator from Declination (latitude) +50° to –50°. There is a generous overlap between them, so it will be easy to find the connection (right).

CHOOSING A STAR MAP

Firstly you need to establish your own latitude so that you can find the corresponding part of the celestial sphere above your head. The circular bands radiating outwards from the poles in Maps 1 and 5 and the horizontal lines across Maps 2, 3 and 4 are lines of Declination. The months along the bottom of the equatorial maps are relevant for observers in the northern hemisphere and those at the top of the page for southern hemisphere observers. If you know your latitude, then the relevant celestial pole will be at your latitude (Declination) above your relevant horizon. Thus at 30°N, the north celestial pole will be at 30° above your northern horizon, and at 30°S, the south celestial pole is 30° above your southern horizon.

Directly above your head, your zenith, is 90° away from your north point in the northern hemisphere and 90° away from your south point in the southern hemisphere. Thus, if your latitude is 30°N, your zenith will be equivalent to Declination +30°, and at latitude 30°S, your zenith is at Declination –30°. If you are at the equator, your zenith is the celestial equator and if you are at the poles, your zenith is the relevant pole at + or –90°.

The Pleiades, a young star cluster, is visible with the naked eye as a close grouping of stars clothed in a faint blue reflection nebula. The individual stars can be distinguished more easily through a telescope.

As the Earth orbits the Sun, the area of sky visible from your latitude varies throughout the year. To discover which constellations are overhead tonight, find the Declination line equivalent to your latitude, find the correct month and the stars shown should be overhead at about 10pm local time. (This does not take into account any daylight saving time.) For example, London at latitude 51°N will have the constellation of Cassiopeia overhead at 10pm at the end of October/beginning of November. For Sydney, at latitude –34°, the sparse constellation of Sculptor is overhead at the same time of year.

FOLLOWING THE PATH OF THE STARS

Once you have established which map is relevant to your sky, you need to consider what position the stars will be in at the time you are observing them. The stars will wheel overhead throughout the night because of the Earth's rotation on its axis. Constellations rise and set in the same manner as the Sun. Therefore, stars rise in the east, reach their highest point at midnight and set in the west. For our equatorial maps the stars move from left to right across the page.

KEY TO STAR MAPS

- All stars of magnitude +6.0 and above are plotted, with their brightnesses shown in half-magnitude steps (right). The larger the circle, the brighter the star.
- All stars, except the faintest, are printed in their spectral colours.
- Double and variable stars are shown, as well as star clusters, nebulae and galaxies.
- The constellation boundaries are shown as broken blue lines; and the internationally accepted asterisms for the constellations in light blue.
- Right Ascension and Declination are shown as a blue grid; the celestial equator is a thicker blue line.
- The Ecliptic is marked in red.
- The names and numbers of the stars are from the Bayer, Flamsteed and Argelander catalogues.
- Deep sky objects are mostly visible only with binoculars or a small telescope. Those from the Messier Catalogue are labelled with the number and the prefix M (for example, the Orion nebula is M42). Objects from the NGC (New General Catalogue) by Dreyer are labelled with just the number. Objects from the IC (Index Catalogue), also by Dreyer, have the prefix I.
- Some of the most interesting objects that appear on each map are mentioned in the descriptions.

JULY

Right: The star maps on the following spreads fit together as shown. Map 1 is the northern circumpolar area and 5 is the southern circumpolar area. Maps 2, 3 and 4 show the equatorial band from 50° north to 50° south. Map 5 is the southern circumpolar area.

MAGNITUDES

		RANGE
-1	○	brighter than -0.5
0	○	-0.5 – 0.5
1	○	0.6 – 1.0
	○	1.1 – 1.5
2	○	1.6 – 2.0
	○	2.1 – 2.5
3	○	2.6 – 3.0
	○	3.1 – 3.5
4	○	3.6 – 4.0
	○	4.1 – 4.5
5	○	4.6 – 5.0
	○	5.1 – 5.5
6	○	5.6 – 6.0

DOUBLE STARS

VARIABLE STARS

SPECTRAL TYPES

O–B A F G K M

OPEN STAR CLUSTERS

GLOBULAR STAR CLUSTERS

PLANETARY NEBULAE

DIFFUSE NEBULAE

GALAXIES

MILKY WAY

MAP ONE

NORTHERN CIRCUMPOLAR REGION

The area shown on this map is always in the northern part of the heavens. The celestial north pole (+90°) is in the centre of the map, and is not visible to people living south of Earth's equator. Because of the Earth's rotation, the sky seems to turn anti-clockwise around this point.

Compared to the southern sky, the northern polar area looks a little empty, but it is worth using a good pair of binoculars to look at the Milky Way area from Perseus to Cygnus. You will be able to see several open star clusters in this area and the easiest target is the Double Cluster in Perseus (NGC869 and 884), close to the boundary with Cassiopeia. The two open clusters, very close together, form a beautiful showpiece especially in a small telescope.

Although they are not as impressive as the Double Cluster, it is worth looking for the three bright star clusters in Auriga, M38, M36 and M37.

An interesting variable star is δ Cephei, close to the Lacerta boundary. It changes from magnitude +3.6 to +4.3 and back again in little more than five days, and is the star that the Cepheid variables are named after. Not far from δ Cephei is another variable star, μ Cephei. It varies between magnitudes +3.6 and +5.1, and is an example of a semiregular variable. In a pair of binoculars you will note the striking red colour of the star. William Herschel called it the 'Garnet Star' and it is one of the reddest stars in the sky.

(continued opposite)

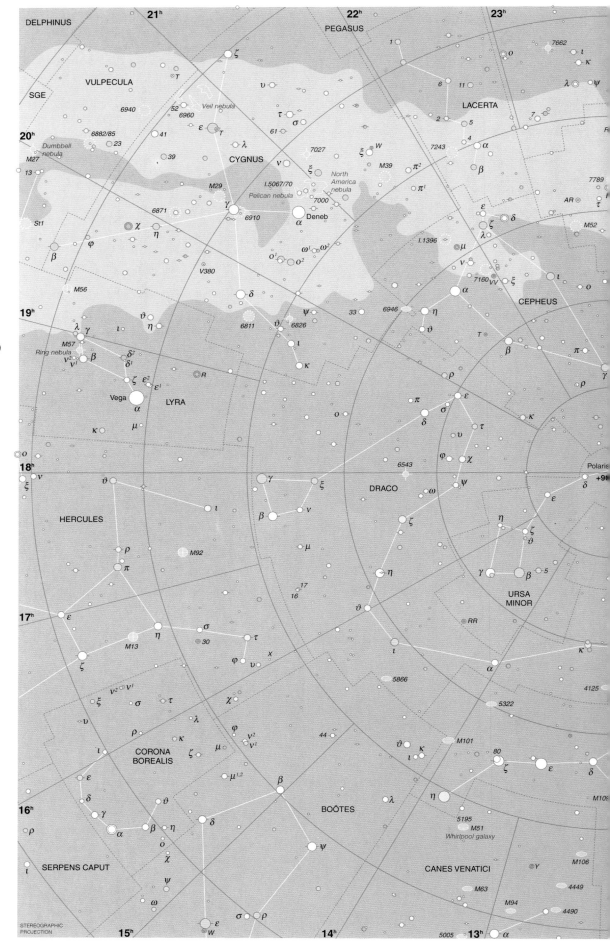

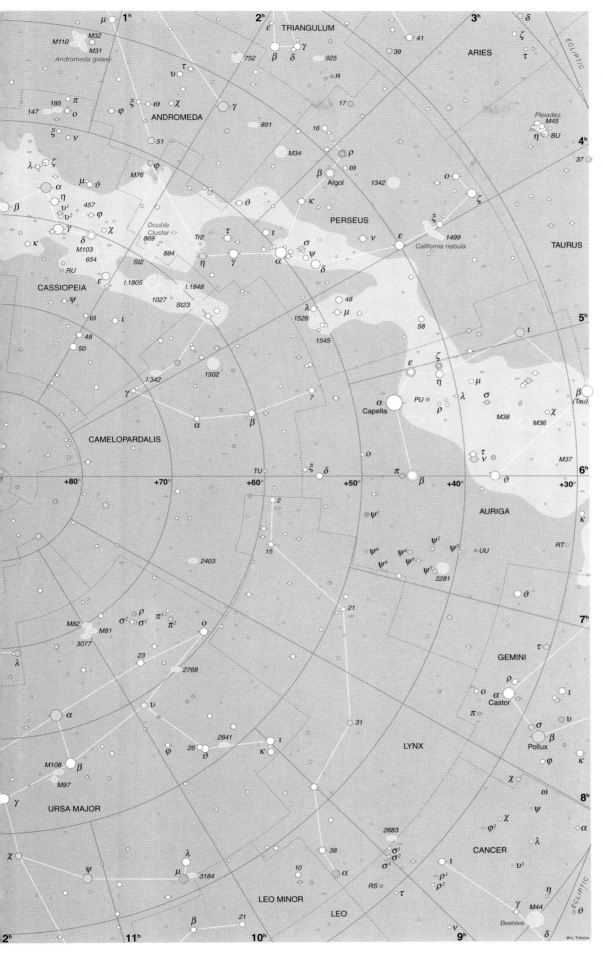

Have a look at β Cygni, or Albireo, in the head of the Swan (the constellation of Cygnus). It is a double star and you will see that the two components have a beautiful contrast in colour. The brighter is goldish-yellow and the fainter component is blue. The two stars can be separated in a good pair of binoculars, but they are more impressive in a small telescope.

One of the best-known double stars is the pair Mizar (ζ) and Alcor (80) in Ursa Major. People with reasonably good eyesight will be able to see both stars with the naked eye, because they are well separated, but the stars do not really form a binary star. They just happen to lie in the same direction as viewed from Earth. A small telescope will show that Mizar itself is a double star. The star has a 4th magnitude companion at a distance of 14.4 arc seconds.

In the head of Draco, the Dragon, you will find an interesting 'equal magnitude' double star. It is ν Draconis, sometimes called Kuma. Both components have a brightness of +4.9 and can easily be separated, even in a small pair of binoculars.

Close to the upper edge of the map is the brightest galaxy in the sky, the Andromeda galaxy, M31, with its two much fainter companions, M32 and M110. Under very good conditions, you can see the Andromeda galaxy with the naked eye, but it will be easier to spot it in a pair of binoculars.

More challenging are the galaxies in Ursa Major, near the bottom of the map. A few interesting targets for the small telescope are the couple M81/M82 (in the upper right part of the constellation), M101 (left, near the border with Boötes) and M51, the Whirlpool galaxy, just below the boundary with Canes Venatici.

MAP TWO

MONOCEROS TO ANDROMEDA

The Andromeda galaxy (see also map 1) will be high in the south in the autumn sky for northern hemisphere observers. A little lower and to the east is another close galaxy, M33, in Triangulum. This is much harder to see than the Andromeda galaxy (M31), as it is much less bright and needs a medium-sized amateur telescope.

In this region of the sky are two well-known variable stars: Mira (o Ceti) and Algol (β Persei). Each of these is the 'prototype' star for a group of variables: Mira for the long-period Mira-type variables and Algol for the Algol-type eclipsing binaries. You should realize that most of the time Mira is faint and even invisible to the naked eye. You will need a small telescope and a detailed star atlas to find the star.

A little farther to the east, in the upper west corner of Taurus, the Bull, is one of the most impressive clusters for a pair of binoculars and even for the unaided eye: M45 or the Pleiades, sometimes called the Seven Sisters.

Not very far from the Pleiades is another, even larger, cluster, the Hyades, with its distinctive 'V' shape. The bright orange star, Aldebaran (α Tauri), the eye of the Bull, is not a member of the cluster, but just happens to be in front of it.

Close to Aldebaran are two naked-eye double stars. A little to the east (left) is σ Tauri, and to the west you will find ϑ Tauri.

Many more open clusters, although less prominent, are found *(continued opposite)*

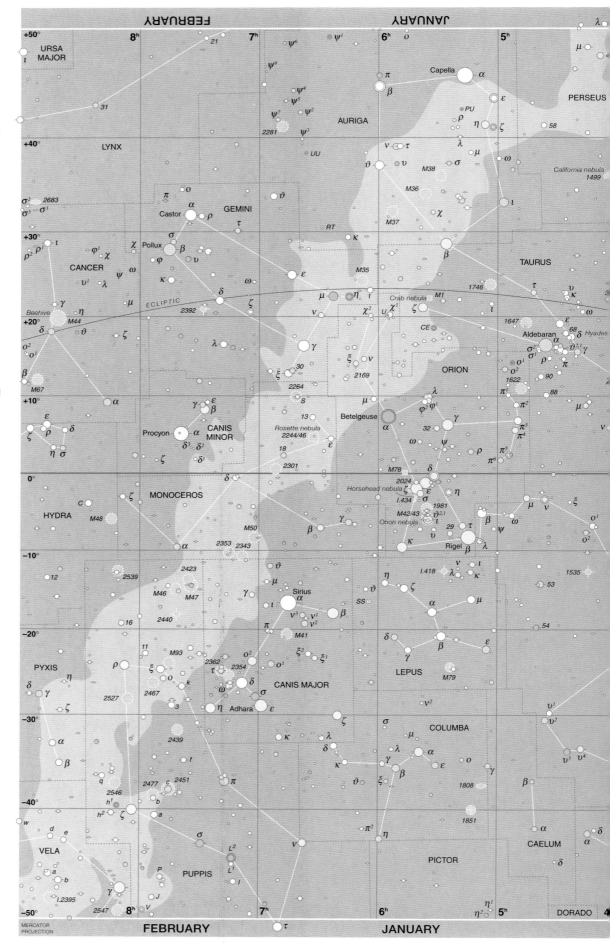

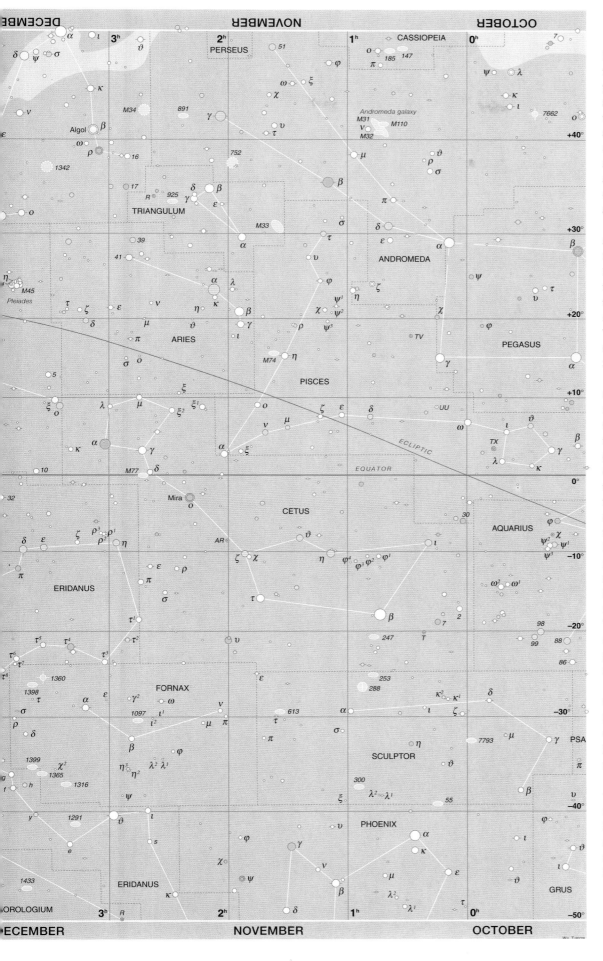

in the Milky Way, from Auriga down to Puppis. In Gemini, close to η, you will find the large cluster M35, and much more to the south you will be able to see M41, situated just a few degrees below the bright star, Sirius. Although these clusters, and several others in the Milky Way area are visible to the naked eye under dark skies, you will usually need a pair of binoculars to find them.

Sirius (α Canis Majoris), sometimes called the Dog Star, lies in the constellation of Canis Major, the Hunting Dog, which follows at Orion's heels. Sirius is the brightest star in the sky and shines with a magnitude of −1.5. The second brightest star is Canopus, (α Carinae) in Carina 35° south of Sirius (see map 5).

One of the most beautiful showpieces in the sky is found in the constellation Orion. It is the Orion nebula, M42, and connected to it is the smaller M43. With good eyesight and in reasonable conditions, you can easily see M42 with the naked eye, just below the three-star belt of Orion, but in a pair of binoculars or, even better, a small telescope, it becomes a real jewel. It is a large cloud of gas and some parts of it are illuminated by bright stars, like mist that is lit up by the headlights of a car. This type of nebula is called a reflection nebula. The majority of the cloud is an emission nebula, meaning that the gas itself is glowing, heated up by nearby stars.

Another emission nebula is the beautiful Rosette nebula (NGC2237) that surrounds the star cluster NGC2244 in Monoceros, just east of Orion. It can be seen in a small telescope.

Back in Orion, it is interesting to compare Betelgeuse and Rigel (α and β Orionis). These two very bright stars have very different colours: Betelgeuse is a red supergiant and Rigel is a hot blue supergiant.

MAP THREE

CORONA BOREALIS TO MONOCEROS

February and March are the best months to watch M44, Praesepe (the Beehive), the open cluster in Cancer, positioned halfway between (and a little to the west of) the two central stars of the constellation. These stars are γ Cancri, also named Asellus Borealis, and δ Cancri, also named Asellus Australis. Under good conditions it is possible to see the Beehive cluster with the unaided eye, but in small binoculars it is really beautiful. Approximately 9° to the south, close to α Cancri, also named Acubens, you will see another open cluster, though much smaller than the Beehive. It is M67.

South of the head of the constellation, Hydra, the Water Snake, and close to the Monoceros boundary is another cluster from the Messier list, M48. Going farther in the same direction, south-west, you will run into another pair of Messier clusters in the constellation Puppis: M46 and the brighter M47.

In the constellations Coma Berenices and Virgo is a very large cluster of galaxies. There are hundreds of galaxies in this area, but on the map only the brightest and best known are plotted. In a small telescope you will not usually see much more than small elliptical, or sometimes circular, patches of faint light. For most of these distant objects a larger telescope is desirable, but you will be able to spot a few. In the centre of the cluster, near the boundary of Coma Berenices and Virgo, you will be able to see two relatively bright galaxies, almost side by side: M84 and M86. *(continued opposite)*

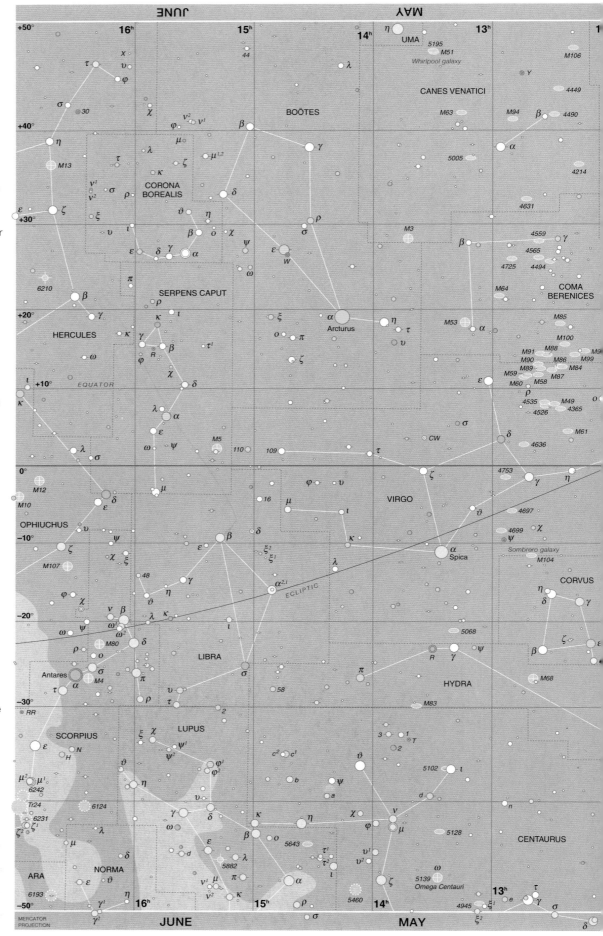

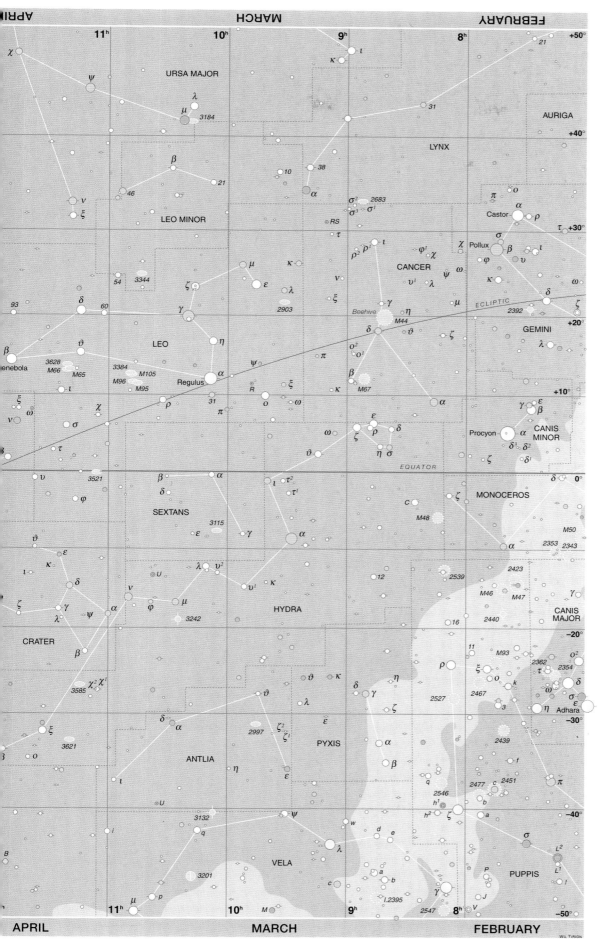

More to the south, and almost at the boundary between Virgo and Corvus, is an interesting galaxy called the Sombrero galaxy, M104. In a small telescope you will see it as a hazy spot with a dark lane in front of it. Photographs reveal the true beauty of the galaxy. We see it almost edge-on and the dark 'lane' is caused by dark nebulosity along the edge of the galaxy.

More to the north and a little to the east is M53, a globular cluster close to the 4th magnitude star α Comae Berenices. Globular clusters form a halo around the central bulge of the galaxy and most of them are found in this area of the sky. The most beautiful globular cluster is Omega Centauri, close to the bottom of the map. In a small telescope it will be resolved into thousands of stars, and is one of the most spectacular southern sky objects.

A few other globular clusters in this area of the sky are M5 in Serpens, M68 in Hydra (just below Corvus), M80 in the head of the Scorpion (Scorpius) and, last but not least, M4. M4 is a beautiful globular cluster and situated in a very interesting area of the sky, near the bright orange star Antares and many bright and dark nebulae, although most of these are not easy to see in a telescope. Photographs will show their real beauty.

West of Scorpius is the constellation Libra (the Scales). α Librae is a naked-eye double star and a good test for your eyesight: if you can separate the two elements, you have good eyesight.

North of Libra is Serpens Caput (the head of the Snake). Serpens is the only constellation in the sky that is divided into two parts, split by the constellation Ophiuchus. Serpens Cauda (the tail of the Snake) can be found on Map 4.

MAP FOUR

ANDROMEDA TO CORONA BOREALIS

The centre of the galaxy is in the western corner of Sagittarius, surrounded by numerous bright objects: nebulae, and open and globular clusters. Several of these objects are visible to the naked eye.

For northern hemisphere observers, examining the Milky Way from the small constellation Scutum, down to the 'tail' of Scorpius, the Scorpion, is a great experience, especially with a good pair of binoculars. You will run into bright open clusters like M11 in Scutum and M24 in Sagittarius, a particularly beautiful and bright section of the Milky Way's star clouds. Farther south you will see the very bright open cluster, M7, in Scorpius, and, in the Scorpion's tail, just north of the naked-eye double star ζ Scorpii, is another rich open cluster, NGC6231. North of it is a rich Milky Way field, where several other clusters can be identified.

There are also several bright gaseous nebulae in this region. The Eagle nebula (M16) is in Serpens Cauda and close to it, the Omega nebula (M17) is in the northern part of Sagittarius. Both are emission nebulae. Closer to the galactic centre are two more beautiful emission nebulae, the Triffid and Lagoon nebulae, M20 and M8.

This is also the realm of the globular clusters. Almost as beautiful as M13 in Hercules is M22, in the centre of Sagittarius. It is easy to *(continued opposite)*

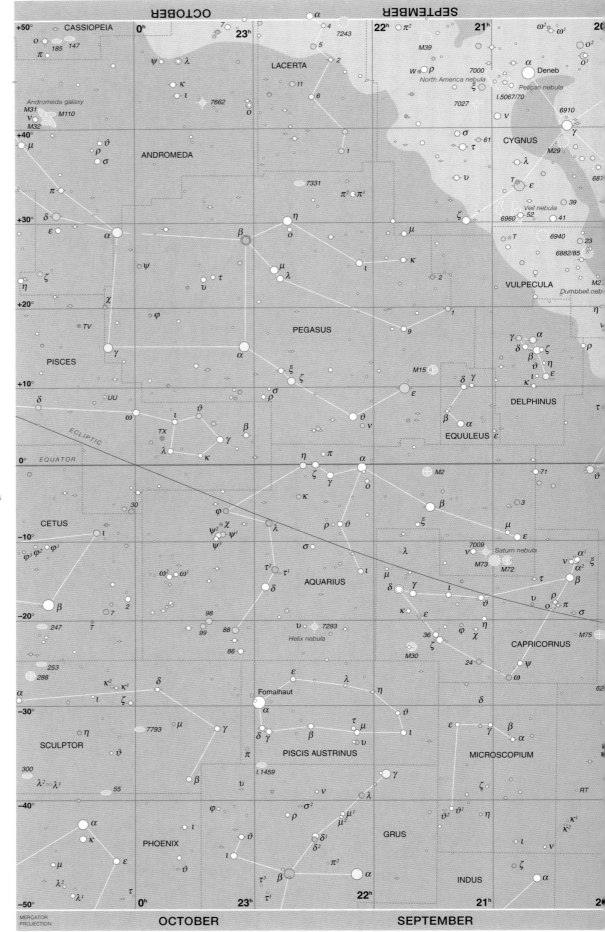

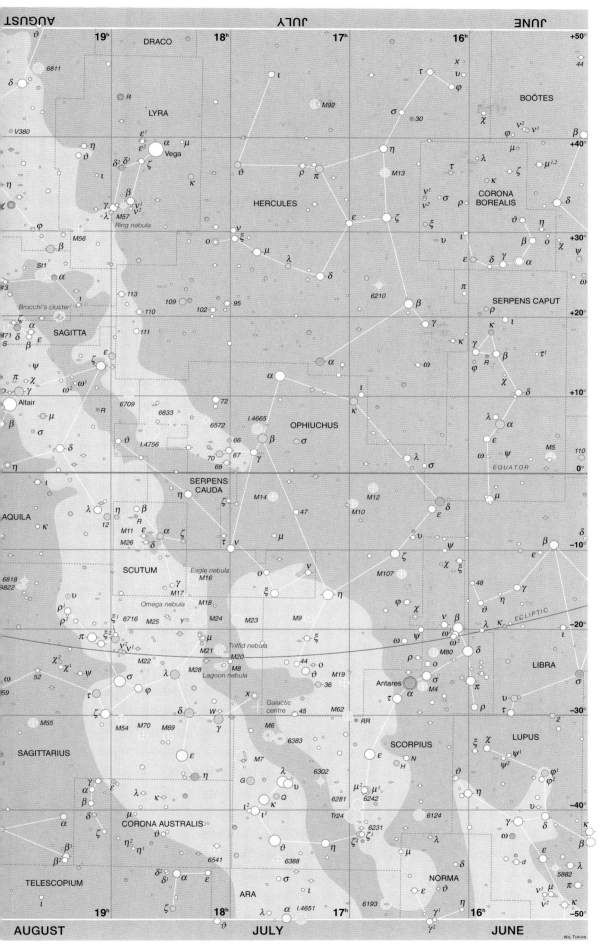

find, since it is surrounded by several bright stars. Another globular cluster, a little more difficult to locate, is M55, which is a little farther to the south-east. When hunting for globular clusters, do not forget M15 in Pegasus, M71 in Sagitta and M2 in Aquarius.

Higher in the sky for northern hemisphere observers is the bright star Vega (α Lyrae), and close to it is a double star, ε Lyrae. Some people with very good eyesight can see the two components without any optical aid, but most people will need a pair of binoculars. The two stars of the pair are almost equal in brightness, and larger telescopes will reveal that each of the two stars itself is double.

A little farther to the south is the 3rd magnitude variable star β Lyrae, which is a close double; the variations in brightness of β Lyrae-type variables are the result of mutual eclipses.

Halfway between β and γ Lyrae is what looks like a small 'unsharp' star. You will need a small telescope to see it and higher magnifications will show that it is a ring. This is M57, the Ring nebula – one of the best-known planetary nebulae. The star is still visible in the centre of the nebula, but it is very faint (14th magnitude), so a large telescope is needed to see it.

Another planetary nebula, and easier to find because it is larger than the Ring nebula, is in Vulpecula. It is M27, the famous 'Dumbbell nebula'. An even larger planetary nebula, but also fainter, is found in the southern part of Aquarius. It is NGC7293, the Helix nebula.

The Veil nebula in Cygnus is a supernova remnant, forming part of what is sometimes called the Cygnus Loop (NGC6960). The nebula is pretty large but also very faint. It is not easy to see in a small telescope, but does show up nicely in good amateur photographs.

MAP FIVE

SOUTHERN CIRCUMPOLAR REGION

The area shown on this map is always in the southern part of the heavens. The celestial south pole (–90°) is in the centre of this map, and is not visible to people living north of the equator. Because of the Earth's rotation, the sky appears to turn clockwise around this point.

Two very prominent objects in this southernmost part of the sky are the Large and the Small Magellanic clouds (LMC and SMC), named after Fernão de Magalhães, a Portuguese explorer who first observed them in the 15th century. These clouds are, in fact, irregular galaxies, companions to the Milky Way.

The Large Magellanic cloud contains one of the brightest emission nebulae in the sky, the Tarantula nebula, also known as 30 Doradus. It is the largest nebula known. If it was at the same distance as the Orion nebula, it would measure 30° across and cover the entire constellation.

Close to the Small Magellanic cloud are two globular clusters, NGC362 and 104. The latter one is also known as 47 Tucanae and is almost as impressive as Omega Centauri (see map 3).

A very bright Milky Way area is found in the constellations of Carina, the Keel, and Crux, the Southern Cross. Around the star ϑ Carinae is a beautiful bright star cluster. It is easily visible to the naked eye and sometimes called the 'Southern Pleiades'.
(continued opposite)

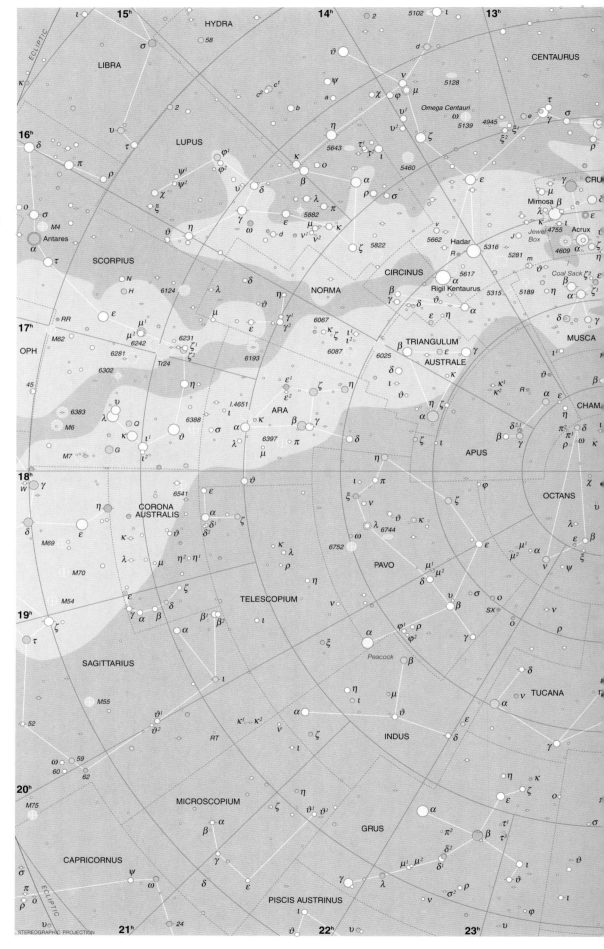

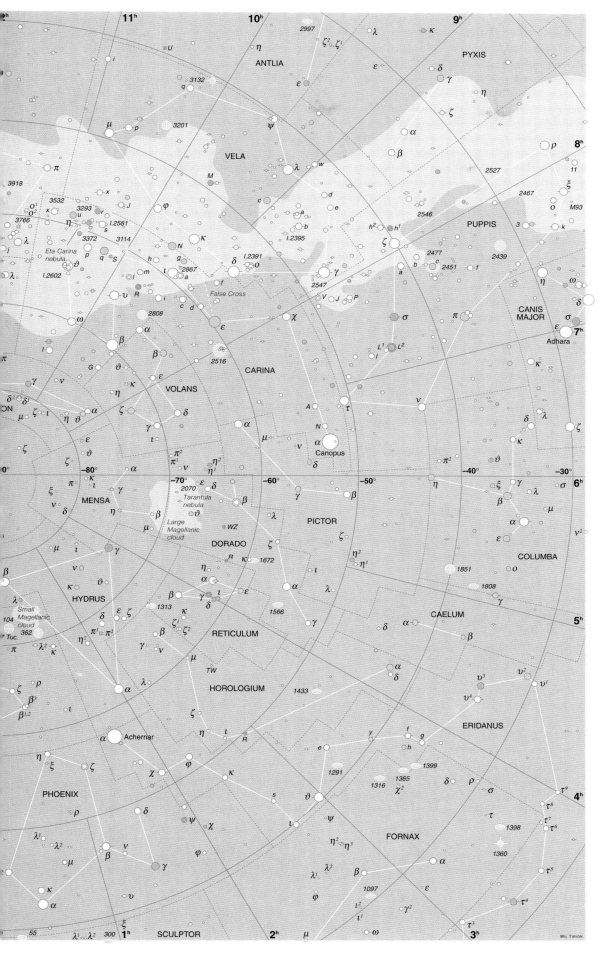

A few degrees to the north is the eta Carina nebula, another naked-eye object, also known as the Keyhole nebula. It contains the very peculiar variable star η (eta) Carinae, which was first recorded as a 4th magnitude star by Edmund Halley in 1730. It varied in brightness irregularly and even reached magnitude –0.8 in 1843, but after that it became invisible to the naked eye.

Close to the Eta Carina nebula are several open clusters. Have a look at NGC3532. It is one of the finest clusters in the Milky Way and has a strange elongated shape.

Crux, the Southern Cross, is without any doubt the best-known constellation in the southern sky and, although small, it contains a few interesting objects. Close to Mimosa (β Crucis), is a very attractive cluster around the 6th magnitude star κ Cru. The cluster NGC4775 is called the Jewel Box.

South of the Jewel Box is a dark area in the Milky Way: the Coal Sack, a cloud of dust about 500 light years away that obscures what lies behind it.

The star α Centauri, or Rigil Kentaurus, is the brightest star in Centaurus. It is a triple star 4.3 light years away. One of the components is Proxima Centauri: this faint 11th magnitude companion, which is almost 2° south of the main pair, is the nearest star to our solar system, but needs a small telescope to see it.

Another interesting star in the multiple star γ Velorum. There are two bright components, magnitudes +1.9 and +4.2, easily seen in a pair of binoculars, and two fainter ones, for which you will need a small telescope. The magnitudes are +8.2 and +9.1.

The brightest of the four components is a very peculiar star. It is the brightest known 'Wolf-Rayet' star. This is a rare class of star, with very high surface temperatures, that seem to eject gas at intervals.

MOON: NEAR SIDE

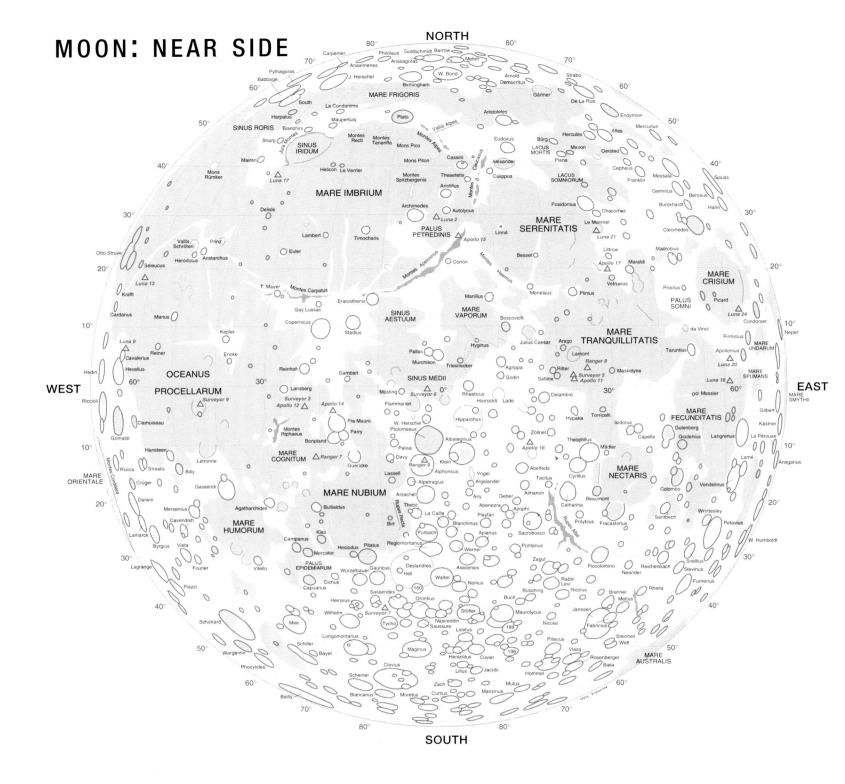

Because of the Moon's synchronous rotation the same side always faces us. Observers using binoculars can see the slight changes due to libration by noting the apparent position of the dark Mare Crisium basin: sometimes it appears very close to the eastern edge of the Moon's visible disk; at other times it is farther away from the limb.

THE FAR SIDE

In October 1959, the Luna 3 probe looped around the Moon and we had our first view of the far side. This contrasts markedly with the near side, having no major dark maria; instead, the terrain is dominated by craters.

THE MOVING TERMINATOR

The features shown in the near-side map are well within reach of amateur instruments. Craters are seen to best advantage when they are close to the day–night line – the terminator. It can be fascinating to follow the terminator's progress across the disk from new moon to full as the Moon waxes, then view the same features under their evening illumination as the Moon wanes, when the sunlight falls from the other side. When the Moon is four to five days old the terminator lies across Theophilus, Cyrillus and Catharina, north of Mare Nectaris. The contrast between the lava plain of Mare Serenitatis and the rugged, cratered south is obvious.

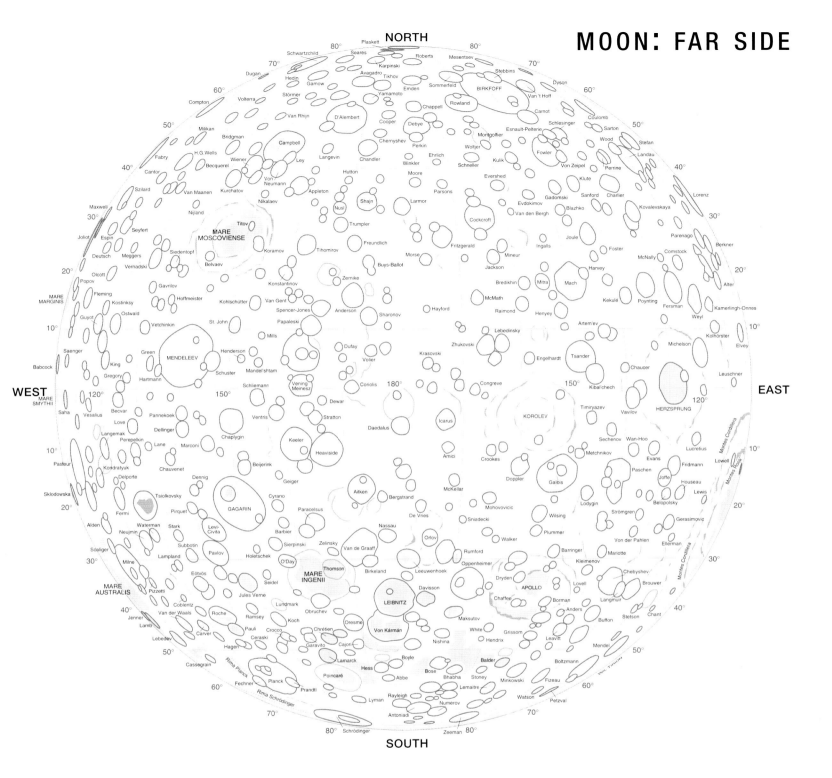

FIRST QUARTER

At first quarter (seven days old), the terminator runs across Ptolemaeus, Alphonsus and Arzachel, near the centre of the visible side of the Moon. These broad, fairly shallow craters have ridged walls that throw long shadows across their floors at local sunrise.

A day later, the huge 225km diameter crater Clavius is in view. Clavius has numerous smaller craters on its floor.

At nine days, the terminator reaches Copernicus. This crater has mountain peaks in its floor, and its appearance changes hourly as the Sun rises. Also in sunlight is dark, lava-filled Plato, just north of the 1300km Mare Imbrium.

At 10 days, the half-drowned crater of Sinus Iridum – a bay on the west of Mare Imbrium – appears to jut beyond the terminator, as the rising Sun catches its western ramparts.

BRIGHT ARISTARCHUS

A day later, Aristarchus, the Moon's brightest feature, moves into view, along with Kepler, another ray centre. At full moon (14 days) shadows are virtually absent, but the contrast between the bright highland regions and the dark lowland maria is at its greatest. This is the best time to see the bright streaks of the ray systems emanating from such comparatively young craters as Tycho and Copernicus.

GLOSSARY

References in *italics* are to other entries within the glossary. References in CAPITALS are to pages 10–11.

Absolute magnitude To compare the intrinsic brightness of stars, astronomers define the absolute magnitude as the *apparent magnitude* the stars would have if observed from a standard distance of 10 PARSECS.

Absorption spectrum A *spectrum* containing dark lines seen against a bright continuous background. In the case of a star these dark lines occur because the atoms in the star's *atmosphere* absorb PHOTONS of characteristic wavelength.

Acceleration Rate of change of *velocity* with time.

Accretion disk A disk of material that orbits an object such as a *black hole* or a star in a close *binary* system. The matter in the disk spirals in toward the object.

Albedo The reflecting power of a planet or other non-luminous body, measured as a percentage or on a scale of 0–1: 0 is totally black; 1 is totally reflecting.

Altazimuth mounting A telescope mounting in which the telescope may move independently about both its horizontal and vertical axes.

Angström unit The unit used to measure the wavelength of light and other *electromagnetic radiation*. It is equal to one ten millionth of a millimetre.

Angular momentum The *momentum* of a body caused by its circular motion around an *axis of rotation*.

Angular velocity The *velocity* of a body rotating about a fixed point.

Antenna A conductor, or system of conductors, for transmitting or receiving *radio waves*. When used in radio astronomy, they are called radio telescopes.

Aperture The clear diameter of a telescope's main mirror or lens.

Aphelion The position of a planet or other body when at the farthest point in its *orbit* from the Sun.

Apparent magnitude The brightness of a celestial object as seen from the Earth.

Atmosphere The gaseous mantle surrounding a planet or other body.

Aurora Aurorae, or polar lights, occur at a height of about 80–100km and are the result of the ionization of the Earth's atmosphere in that region. They are caused by the interaction of particles from the *solar wind* and trapped particles in the Earth's *Van Allen belts*. Aurorae have also been seen on Jupiter and Saturn.

Axis of rotation Imaginary line about which a planet, star or other celestial body rotates.

Bar Unit of pressure. One bar is equal to the pressure of the Earth's *atmosphere* at sea level.

Barred spiral galaxy A galaxy with spiral arms emerging from either end of a bar that crosses the galactic plane.

Basalt A type of *igneous* rock.

Big Bang The beginning of the universe.

Binary star A system of two stars that are moving round their common centre of gravity.

Black hole A massive object so dense that no light or other radiation can escape from it.

Blazar A *quasar* with distinctive, extreme characteristics.

Brown dwarf A 'star' that never became large enough for the pressure and temperature at its core to start nuclear fusion.

Cassegrain reflector A *reflecting telescope* in which the light from the object under study is reflected from the main mirror to a convex secondary mirror, and from there back to the eyepiece through a hole in the main mirror.

Cepheid variable star Cepheids vary over periods of between a few days and a few weeks. They are regular in their behaviour. It has been found that the period of a Cepheid is linked with its *absolute luminosity*: the longer the period, the more luminous the star. From this, it follows that once a Cepheid's period has been measured, its luminosity can be calculated and its distance determined.

Chromosphere The part of the Sun's atmosphere above the *photosphere*, and below the *corona*. It is visible to the naked eye during a total *solar eclipse*.

Circumpolar star A star that never sets, but which circles the CELESTIAL POLE and remains above the horizon for an appropriately paced observer.

Clusters See *globular clusters, open clusters*.

Colour index The difference of a star's *apparent magnitude* measured in two standard wavelengths. It is a measure of a star's colour, and so of its surface temperature.

Coma The bright nebulous cloud of dust and gas that surrounds the nucleus of a COMET when it is close to the Sun.

Conjunction The apparent close approach of a planet to a star or another planet; it is purely a line-of-sight effect. See *inferior conjunction, superior conjunction*.

Constellation A group of stars named after a living or a mythological character, or an animal or inanimate object. The stars in a constellation are not genuinely associated with each other; they lie at very different distances from the Earth, and simply happen to be in roughly the same direction in space.

Corona The outermost part of the Sun's *atmosphere*. Its nature varies with the *sunspot* cycle, and it is visible with the naked eye during *solar eclipses*.

Cosmic rays Highly energetic particles that move through space at velocities close to the speed of light. Some enter the Earth's *atmosphere*.

Cosmology The study of the universe as a whole.

Culmination The time when a star or other body is at its highest point in the sky, when it reaches the observer's MERIDIAN.

Day The time taken for a planet to spin once on its axis.

Density The mass of a given substance per unit of volume. If the density of water is taken as 1, the Earth's density is 5.5.

Dobsonian mount A simple, stable telescope mounting in which the telescope may move independently about both its horizontal and vertical axes.

Doppler effect The apparent change in the wavelength of light caused by the relative motion of it and the observer.

Double star A star that appears to be made up of two components when viewed from Earth. Some doubles are optical; that is to say, the components are not truly associated, and simply happen to lie close together as seen from Earth. Some double stars, however, are physically associated or *binary* systems.

Earthshine The faint luminosity of the night hemisphere of the Moon, caused by light reflected onto the Moon from the Earth.

Eccentricity The extent to which the *orbit* of a planet or other body varies from a perfect circle.

Eclipses, See *lunar eclipse, solar eclipse*

Electromagnetic radiation Energy in the form of electromagnetic waves.

Electromagnetic spectrum The full range of *electromagnetic radiation*: *gamma rays, X-rays, ultraviolet, visible light, infrared* and *radio waves*. Visible light makes up only a very small part of the whole electromagnetic spectrum.

Ellipse Any section cut through a cone at an angle that does not cut the base of the cone is an ellipse. Most planets and other bodies in the solar system have elliptical rather than circular *orbits*.

Elliptical galaxy A galaxy with no central bulge or spiral arms. They can vary in shape from almost spherical to elongated.

Elongation The apparent angular distance of a planet from the Sun, or of a satellite from its parent planet.

Emission spectrum A *spectrum* containing bright lines or bands produced by the emission of PHOTONS.

Equation of time Because the Earth's *orbit* is not circular, the Sun does not appear to move across the sky at a constant rate. The equation of time is the difference between the apparent solar time and mean solar time. It is at its greatest during the months of February and November.

Equator An imaginary line drawn on a planet or other body with all points on it at an equal distance from the north and south poles.

Equatorial mounting A telescope mounting in which the instrument is set upon an axis that is parallel to the Earth's axis. This has the benefit for observers that to keep an object in view the telescope need be turned only in RIGHT ASCENSION.

Equinoxes The vernal (March 21) and autumnal (September 21) equinoxes are the dates when the Sun crosses the CELESTIAL EQUATOR and day and night are of equal duration.

eV Electron volts are units of energy. One electron volt is equivalent to the energy of one electron falling through one volt.

Eyepiece A system of lenses attached to a telescope and through which the observer looks. The magnifying power of the telescope can be altered by changing the eyepiece.

Faculae Bright, temporary patches in the upper of the Sun that are usually (not always) associated with *sunspots*.

Finder A low-power telescope attached to a larger one and aligned with it. Its wider field of view makes it useful for locating celestial objects that can then be looked at in detail through the main telescope.

First Point of Aries The point on the sky from where RIGHT ASCENSION is measured. It is the position of the Sun at the vernal *equinox*.

Flares, solar Explosive releases of energy in the outer part of the Sun's *atmosphere*, often associated with active sunspots. They send out electrified particles that may reach the Earth, causing magnetic storms and *aurorae*.

Focal length The distance between a lens or mirror in a telescope or binoculars and the point at which the image of an object at infinity is focused.

Fraunhofer lines The dark *absorption* lines in the Sun's spectrum. See *spectrum, stellar*.

Gamma rays Extremely short-wavelength *electromagnetic radiation*.

Gas giant The large outer planets of the solar system: Jupiter, Saturn, Uranus and Neptune.

Gegenschein A very faint glow in the sky, exactly opposite to the Sun. Part of the *Zodiacal light*, it is very difficult to observe.

Gibbous A phase of the Moon or planet that is more than half, but less than full.

Globular clusters Collections of tens of thousands of stars. Roughly spherical in shape, they are very remote and lie in a halo around the central bulge of the *Milky Way* and other *spiral* and irregular galaxies.

Gravitation The force of attraction that exists between all particles of matter in the universe.

Gravitational lens A galaxy or other massive object whose gravity bends the light from a more distant object, creating a magnified image of it.

Great Circle A circle on the surface of a sphere (such as the Earth or the CELESTIAL SPHERE): its plane passes through the centre of the sphere.

H-I regions Clouds of neutral hydrogen that cannot be seen, but may be studied by radio telescopes.

H-II regions Clouds of hydrogen ionized (usually) by the presence of hot stars. The recombination of ions and electrons leads to the emission of light, by which the H-II regions can be seen.

Hertzsprung-Russell diagram A diagram in which stars are plotted according to their *spectral* type (or temperature) and luminosity. See *spectrum, stellar*.

Hour angle The time that has elapsed since a celestial body crossed the observer's MERIDIAN. Zero hour is when an object is on the observer's meridian.

Hour circle A great circle on the CELESTIAL SPHERE that passes through both CELESTIAL POLES.

Hubble's constant The rate at which the expansion *velocity* of the universe changes with distance.

Hydrogen I region, see *HI regions*.

Hydrogen II region, see *HII regions.*

Igneous rock Rock derived from *magma* or lava that has solidified.

Index catalogue Two supplements to the *New General Catalogue* were published in 1895 and 1908. Objects within these are identified by their IC number.

Inferior conjunction For Mercury and Venus; the position when the planet is in line with the Sun, on the same side as the Earth.

Inferior planet A planet closer to the Sun than the observer. From the Earth, Mercury and Venus are inferior planets.

Infrared radiation *Electromagnetic radiation* with wavelengths longer than *visible light*. We cannot see it, but feel it as heat.

Interstellar medium Dust and gas that lies between the stars of the galaxy.

Interstellar extinction The reduction of brightness in objects due to the absorption and scattering of the object's radiation by the interstellar medium.

KeV Units of energy denoting thousands of *electron volts*.

Librations, lunar Although the Moon's rotation is captured with respect to the Earth, there are effects, known as librations, that enable us to examine 59 per cent of the total surface, though no more than 50 per cent can be seen at one time. There are three librations: in longitude (because the Moon's orbital *velocity* is not constant); in latitude (because the Moon's equator is inclined by 6° to its orbital plane); and diurnal (due to the rotation of the Earth.

Limb The edge of the visible disk of the Sun, Moon or planet.

Local Group of galaxies The group of galaxies of which our galaxy, the *Milky Way*, is a member.

Luminosity The total intrinsic brightness of an object.

Lunar eclipse When the Earth lies directly between the Sun and the Moon, its shadow falls on the lunar surface.

Magma Molten rock.

Magnetic field A field of force around a magnet. The liquid metallic outer core of the Earth acts as a magnet so the Earth has a magnetic field. Stars have magnetic fields that are caused by the movement of charged particles within them.

Magnetosphere The region surrounding a magnet in which its magnetic feld has an effect. The Earth's magnetic field extends out to about 10 Earth radii.

Magnitude See *absolute magnitude*, *apparent magnitude*

Main sequence The band on the *Hertzsprung-Russell diagram* where stars lie for most of their lives, burning the hydrogen in their cores.

Mare Relatively smooth, dark areas on the surface of the Moon or planet.

Mass The quantity of matter that a body contains.

Messier numbers Numbers given by the 18th-century French astronomer and comet hunter, Charles Messier, to various nebulous objects that he did not wish to mistake for comets. His numbers are still used, thus the Andromeda galaxy is M31.

Microwave radiation Electromagnetic radiation with wavelengths of between 1mm and 30cm.

Milky Way A dense band of stars, gas and dust seen from dark sites. It is the central plane of our galaxy. The name given to our galaxy.

Millibar Unit of pressure: one thousandth of a *bar*.

Minor planet Another term for ASTEROID.

Moon The Earth's natural satellite.

Momentum A body's mass multiplied by its *velocity*.

Nadir The point on the CELESTIAL SPHERE directly below the observer.

Neutron star The final stage in a giant star's lifecycle. The force between individual neutrons stops the star from further collapse. Neutron stars are incredibly dense.

New General Catalogue (NGC) Catalogue of objects including nebulae and star clusters drawn up by J L E Dreyer. Objects are often denoted by their NGC number. For example the Andromeda Galaxy is NGC224.

Newtonian reflector A *reflecting telescope* in which light is collected by a curved main mirror, reflected back to a second small, flat mirror, which is at an angle to reflect the light to the side of the telescope's tube, where it is focused.

Noctilucent clouds Clouds that appear to shine at night. They are at a height of about 80km in the Earth's ionosphere and reflect the light of the Sun.

Nodes The points at which the orbit of an object crosses a reference plane. The reference plane is usually the ECLIPTIC or CELESTIAL EQUATOR.

Nova A star that suffers a sudden outburst, and flares up to many times its normal brightness for a short time before fading back.

Nucleus, cometary The rocky and icy centre of a COMET.

Object-glass The main lens of a *refracting telescope*.

Objective The light-gathering lens or mirror in a telescope.

Oblateness The degree of flattening of an spheroid. The Earth is oblate – its diameter at the equator is greater than that from pole to pole.

Occultation The complete or partial covering up of one celestial body by another. Thus the Moon may pass in front of (and occult) a star or planet.

Open clusters A loose collection of, at most, a few thousand stars. Naked-eye examples in include the Pleiades and Hyades.

Opposition The position in a planet's *orbit* when it is exactly opposite the Sun as seen from the Earth.

Orbit The path of one celestial body around another.

Parallax The apparent shift in the position of a body against objects further away when observed from two different places. Astronomers can calculate the distances of nearby stars by measuring the apparent shift due to the Earth's own movement round the Sun.

Penumbra The outer, relatively light parts of a *sunspot*. The area of partial shadow lying to either side of the main cone of shadow cast by an object.

Perihelion The closest approach of a body (such as a planet or comet) to the Sun.

Photosphere The bright 'visible surface' of the Sun.

Planetary nebula An expanding cloud of gas that has been ejected by a star in the later stage of its life.

Population, stellar Two main classes of stars. See *Population I*, *Population II*.

Population I Young, metal-rich luminous stars associated with regions of gas and dust, eg, in the arms of *spiral galaxies*.

Population II Old red stars associated with regions devoid of gas and dust, for example in *elliptical galaxies*.

Position angle The apparent direction of one object with reference to another, measured easterly from the north point of the main object.

Precession The apparent slow movement of the celestial poles over a period of 25,800 years.

Prism A block of glass with flat surfaces inclined to one another. Light passing through the prism will be split up into a spectrum as the different constituent colours of visible light are refracted by different amounts. See *spectrum, visible*.

Prominence Masses of glowing gas, chiefly hydrogen, above the Sun's *photosphere*.

Proper motion The individual motion of a star on the *celestial sphere* due to its own motion.

Protostar A star in the process of formation in which nuclear fusion has not yet commenced.

Pulsar A source that emits *radio waves* in rapid, very regular pulses. They are *neutron stars*.

Quasar (Quasi-stellar object) A very bright, extremely distant object. Although quasars appear starlike in telescopes, they are thought to be the hyperactive centres of galaxies.

Radiation See *electromagnetic radiation*.

Radio galaxy A galaxy that is a very powerful emitter of *radio waves.*

Radio telescope See *antenna*.

Radio waves Part of the *electromagnetic spectrum*, with very long wavelengths.

Red dwarf *Main sequence* stars that are cooler and smaller than the Sun.

Red giant A large reddish or orange star in a late stage of its lifecycle. It is relatively cool and has expanded to perhaps 100 times its original size.

Redshift An apparent shift of spectral lines in the radiation of an object toward the red end of the spectrum, caused by the relative motion of the object away from the observer. See *Doppler effect*.

Reflecting telescope A telescope in which the light is collected by means of a mirror.

Refracting telescope A telescope in which the light is collected by means of a lens.

Retrograde motion The apparent motion from east to west of an object as

seen from Earth. *Superior planets* can experience retrograde motion when the Earth undertakes them in its orbit. In general, objects move with direct motion from west to east.

Roche limit The distance from the centre of a planet or star, within which another body would be broken up by gravitational distortion.

Seasons Effects on the climate due to the inclination of a planet's axis.

Sedimentary rock Rock formed from material deposited by liquid, ice or wind and then compressed.

Seeing The quality of observing conditions. The steadiness and clarity of an object's image, depends upon the amount of turbulence in the Earth's *atmosphere*.

Sidereal period The time taken for a planet or other body to make one *orbit* around the Sun with reference to the stars.

Sidereal time The local time reckoned according to the apparent rotation of the CELESTIAL SPHERE. It is 0 hours when the *First point of Aries* crosses the observer's MERIDIAN.

Singularity A mathematical point at which space and time are infinitely distorted. It is believed that a *black hole* contains a singularity.

Solar eclipse When the Moon passes directly between the Sun and the Earth it casts a shadow on the Earth. An observer from within the shadow will see the *photosphere* blocked by the Moon and will be able to see the Sun's outer *atmosphere*.

Solar system The group of objects containing the Sun and any object gravitationally bound by it. The group includes the planets and their satellites, asteroids, comets, meteors and interplanetary dust and gas.

Solar wind A flow of energetic, charged particles streaming out from the Sun. The wind's *velocity* in the region of the Earth exceeds 950km/sec.

Solstices The dates on which the Sun reaches its northernmost (June 21) and southernmost (December 21) Declinations.

Spectral class Classes in which stars are grouped according to the characteristic emission lines in their spectra. See also *spectrum, stellar*.

Spectrohelioscope An instrument used to view the Sun's radiation at one particular wavelength.

Spectroscope An instrument used to analyse the *spectrum* of a star or other luminous object.

Spectroscopic binary A *binary* star whose components are too close together to be seen separately, but whose relative motions cause *Doppler* shifts that are detectable using a *spectroscope*.

Spectrum, electromagnetic See *electromagnetic spectrum*.

Spectrum, stellar The spectrum of a star has characteristic *absorption* and *emission* lines that can give a great deal of information about the star.

Spectrum, visible The *electromagnetic radiation* that we see as light split into its constituent colours. The best known manifestation of this is a rainbow.

Spiral galaxy A galaxy with well-defined spiral arms that are orbiting around a central nucleus.

Sun The star around which the Earth orbits.

Sunspots Darker patches on the Sun's *photosphere* that are about 2000K cooler than the surrounding area. Large sunspots have a dark central *umbra* and a lighter, outer *penumbra*. They are regions of strong *magnetic fields*, and appear on the Sun in an 11-year cycle.

Supergiants The largest and most luminous type of star.

Superior conjunction The position of a planet when it is on the far side of the Sun in relation to the Earth.

Superior planet A planet farther out from the Sun than the Earth.

Supernova A star that suffers a cataclysmic explosion, sending much of its material into space. There appear to be two types: one class attains a maximum luminosity of 100 million Suns, the other of 10 million Suns. Supernova remnants are strong radio sources.

Terminator The day-night line on a planet or satellite.

Terrestrial planets The four inner planets of the solar system.

Tides Distortion of a planet or star, etc by the effect of gravity from another object or objects. The regular rise and fall of the Earth's ocean waters, caused by the gravitational pull of the Moon and Sun.

Transit The passage of a body between the Earth and the Sun so that it is seen crossing the disk of the sun. The passage of a satellite in front of its parent planet. The passage of a celestial body across the observer's MERIDIAN.

Ultraviolet radiation *Electromagnetic radiation* that has a wavelength just shorter than *visible light*.

Umbra The dark inner portion of a *sunspot*. The main cone of the shadow cast by an object.

Van Allen belts Zones around the Earth in which electrically charged particles are trapped and accelerated by the Earth's *magnetic field*.

Variable stars Stars that fluctuate in brightness over relatively short periods.

Velocity The rate of motion in a given direction.

Visible light The part of the *electromagnetic spectrum* to which our eyes react.

Visual binary A *binary* pair far enough apart to be visible separately.

Weight The force experienced by a body due to the effect of gravity on its mass.

White dwarf The remnants of lower mass stars that are no longer undergoing nuclear fusion. The force between electrons stops the star collapsing further.

X-rays *Electromagnetic radiation* with very short wavelengths.

Zodiac A belt stretching right around the sky about 9° to either side of the *ecliptic*, in which the Sun, Moon and bright planets are always to be found.

Zodiacal light A cone of light rising from the horizon and stretching along the *ecliptic*. It is visible only when the Sun is a little way below the horizon and is due to sunlight being scattered by minute dust particles in the inner solar system.

INDEX

All page references in **bold** refer to photographs, illustrations and boxes. Numbers that appear first in entries are arranged as if spelt out, eg 51 Pegasi is found under F. Similarly, Greek letters are arranged as if spelled out.

ACKNOWLEDGEMENTS

Publisher's acknowledgements
The publishers would like to thank all the contributors to the book, as well as all the individuals and institutions who provided photographs or information.

General Editor's acknowledgements
I would like to thank everyone who has helped and contributed to this book. It has been an immensely interesting and enjoyable task condensing the universe into one volume.

Photographic acknowledgements

Key: NASA HQ = NASA Headquarters; SPL = Science Photo Library; AAO = Anglo-Australian Observatory; BAS = Bridgend Astronomical Society; GSPL = Genesis Space Photo Library; NOAO = National Optical Astronomy Observatories; ROE = Royal Observatory, Edinburgh

Front cover, left NASA HQ, Front cover, right NASA HQ, Front cover top NASA HQ, Back cover, top NASA HQ; Back cover, bottom ROE/AAO From UK Schmidt plates by David Malin.

Endpapers NASA HQ;1 NASA HQ; 2 Corbis UK Ltd/Roger Ressmeyer; 3 NASA HQ; 4–5 ROE/AAO From UK Schmidt plates by David Malin; 6–7 SPL/Frank Zullo; 8 NASA HQ; 9 left Lowell Observatory; 9 top right NASA HQ; 9 centre right NASA HQ; 9 bottom right Corbis UK Ltd/Betmann; 12 bottom SPL/NOAO; 12–13 top SPL/Dr Rudolph Schild; 13 top SPL/Luke Dodd; 13 centre SPL/David A Hardy; 13 bottom NASA HQ; 20 SPL/NOAO; 21 SPL/European Space Agency; 22 top National Solar Observatory/Sacramento Peak; 22 bottom SPL/NOAO; 23 NASA HQ; 24 NASA HQ; 25 NASA HQ;27 NASA HQ; 28 Corbis UK Ltd/NASA-JPL; 29 SPL/Pekka Parviane; 32 Jan Curtis/Geophysical Institute, University of Alaska Fairbanks; 33 Centre Spatial de Toulouse/Dr Massonnet; 34 NASA HQ; 35 NASA HQ; 37 NASA HQ; 38 Dr Martin Wooster/Department of Geography, King's College London; 38–39 bottom Corbis UK Ltd/Bob Krist; 40 top left Martin Griffiths/BAS; 40 top right Martin Griffiths/BAS; 41 top left Martin Griffiths/BAS; 41 top centre left Martin Griffiths/BAS; 41 top centre Martin Griffiths/BAS; 41 top centre right Martin Griffiths/BAS; 41 top right Martin Griffiths/ BAS; 41 bottom NASA HQ; 42 NASA HQ; 43 NASA HQ; 44 top Corbis UK Ltd/Dennis di Cicco; 44 bottom SPL/ Dr Fred Espenak; 45 SPL/Re Ronald Royer; 47 top NASA HQ; 47 bottom NASA HQ;48 left Private Collection; 48 top NASA HQ; 49 top NASA HQ; 49 bottom NASA HQ; 50 left NASA HQ; 50 right NASA HQ; 50 top Finley-Holiday FIlms Stock Photos/JPL; 51 top Finley-Holiday Films Stock Photos/JPL; 51 bottom Finley-Holiday FIlms Stock Photos/JPL; 52 NASA HQ; 52–53 bottom Natural History Museum; 55 NASA HQ; 56 NASA HQ; 57 NASA HQ; 58 top left NASA HQ; 58 top right NASA HQ; 58 bottom left NASA HQ; 59 top right AKG, London/Erich Lessing/Museo delle Scienze – Florence; 59 bottom left NASA HQ; 60 top NASA HQ; 60 bottom NASA HQ; 61 left NASA HQ; 61 right NASA HQ; 62 NASA HQ; 63 NASA HQ; 64 left Private Collection; 64 right NASA HQ; 65 top NASA HQ; 65 bottom left NASA HQ; 65 bottom right NASA HQ; 66 top NASA HQ; 66 bottom left NASA HQ; 66 bottom right NASA HQ; 67 top NASA HQ; 67 bottom left NASA HQ; 67 bottom right NASA HQ; 69 top NASA HQ; 69 bottom NASA HQ; 70 top left NASA HQ; 70 top right NASA HQ; 70 bottom AKG, London; 71 top left NASA HQ; 71 top right NASA HQ; 71 bottom NASA HQ; 72 NASA HQ; 73 NASA HQ; 74 top NASA HQ; 74 bottom NASA HQ; 75 top Hulton Getty Picture Collection; 75 bottom Finley-Holiday FIlms Stock Photos/NASA; 76 Lowell Observatory; 77 top NASA HQ; 77 centre NASA HQ; 77 bottom Corbis/Betmann; 78 left NASA HQ; 78 right NASA HQ; 79 SPL/Pekka Parviainen; 80 top NASA HQ; 80 bottom NASA HQ; 81 top NASA HQ; 81 centre NASA HQ; 81 bottom NASA HQ; 82 top left Hulton Getty Picture Collection; 82 top right NASA HQ; 82 bottom SPL/Gordon Garradd; 84 top Galaxy Picture library/Michael Stecker; 84 bottom NASA HQ; 85 Queen's University of Belfast/A; Fitzsimmons; 86 top Natural History Museum; 86 bottom Royal Astronomical Society/Kitt Peak; 87 top SPL/Francois Gohier; 87 bottom N & S Szymanek; 88 top NASA HQ; 88–89 bottom NASA HQ; 90 Private Collection 91 NASA HQ; 93 NASA HQ; 94 Corbis UK Ltd; 95 National Solar Observatory/Sacramento Peak; 96 top Seth Shostak; 96–97 bottom Seth Shostak; 100 NASA HQ; 101 AAO/David Malin; 102 top NASA HQ; 102 bottom NASA HQ; 103 top left AAO/David Malin; 103 top right AAO/David Malin; 103 bottom NASA HQ; 104 top Royal Astronomical Society; 104 bottom left SPL/NOAO; 104 bottom right SPL/John Sanford; 105 SPL/Pekka Parviainen; 106 top Neil Bone; 106 bottom Neil Bone; 107 NASA HQ; 108 SPL/John Sanford; 109 top left NASA HQ; 109 top right Private Collection; 109 bottom NASA HQ; 112 top Lund Observatory, Sweden; 112 bottom William Herschel; 113 top Isaac Roberts ;113 bottom AAO/David Malin; 114 top AAO/Courtesy Sky Publishing Corp; 114 bottom left AAO/David Malin; 114 bottom right AAO/David Malin; 115 top AAO/David Malin; 115 bottom AAO/David Malin; 116 AAO/David Malin; 117 top AAO/David Malin; 117 bottom AAO/Kitt Peak Observatory; 118 top AAO/David Malin; 118 bottom AAO/David Malin; 119 top AAO/David Malin; 119 bottom AAO/IAC photo taken from plates taken with the Isaac Newton telescope – photo by David Malin; 120 top AAO/David Malin; 120 bottom left AAO/Royal Observatory, Edinburgh from UK Schmidt Telescope plates by David Malin; 120 bottom right AAO/David Malin; 121 top left AAO/ROE from UK Schmidt Telescope Plates by David Malin; 121 top right AAO/David Malin; 121 bottom NASA HQ; 122 top left SPL/Dr Ian Robson; 122 centre left NASA HQ; 122 bottom AAO/Hubble Space Telescope; 126 California Institute of Technology and Carnegie Institute of Washington; 127 Isaac Newton Group of Telescopes/William Herschel Telescopes/British PPARC/Dutch NWO; 128 top Corbis UK Ltd/Michael Yamashita; 128 bottom Corbis UK Ltd/Roger Ressmeyer; 129 NASA HQ; 130 NASA HQ; 131 NASA HQ; 132 top SPL; 132–133 bottom California Institute of Technology and Carnegie Institution of Washington; 134–135 NASA HQ; 136 top Royal Astronomical Society; 136 bottom AAO/David Malin; 137 top European Southern Observatory; 137 bottom Corbis UK Ltd/Roger Ressmeyer; 138 top NOAO/International Dark Sky Association; 138–139 bottom NOAO/International Dark Sky Association; 139 European Southern Observatory/JM Braun, Sternwarte Bonn; 140 left Seth Shostak/SETI Inst; 140 right Seth Shostak; 140 centre GSPL/ESA; 141 right GSPL; 141 top left NASA HQ; 141 top centre left GSPL; 141 top centre right GSPL; 141 bottom Corbis UK Ltd/Roger Ressmeyer; 142–143 top Seth Shostak; 144 right NASA HQ; 144–145 bottom NASA HQ; 146 top Hulton Getty Picture Collection; 146 bottom left NASA HQ; 146 bottom right GSPL; 147 top NASA HQ; 147 bottom Rex Features/Sipa Press; 148 left SPL/Ton Kinsbergen; 148 right NASA HQ; 149 top centre right NASA HQ; 149 top right NASA HQ; 149 bottom left Max-Planck-Institut für Radioastronomie, Bonn; 149 bottom right NASA HQ; 150 top Seth Shostak; 150–151 bottom Seth Shostak; 152–153 Jim Kaler; 154 AAO/David Malin; 155 left Galaxy Picture Library; 155 right Tim Hicks; 157 Pam Spence; 158 top SPL/NASA/GSFC; 158–159 bottom Hansen Planetarium Publications; 160 SPL/Chris Madeley; 161 SPL/Richard J Wainscoat, Peter Arnold Inc; 162 SPL/David Nunjk; 163 Jan Curtis; 164 Astronomy Now; 165 top left N & S Szymanek; 165 top right Astronomy Now; 165 bottom Astronomy Now; 166 AKG, London; 168 right NASA HQ; 168 top left Nik Szymanek/Ian King; 169 top left Nik Szymanek/Ian King; 169 centre left Nik Szymanek/Ian King; 169 bottom left Nik Szymanek/Ian King; 169 bottom right NASA HQ; 170–171 California Institute of Technology and Carnegie Institution of Washington.

Artwork acknowledgements

14–15 all Susanna Addario/Thomas Trojer; 16–17 Mark McLellan; 18 Mark McLellan; 19 Mark McLellan; 20 Hardlines; 21 Susanna Addario/Thomas Trojer; 23 Susanna Addario/Thomas Trojer; 24 left Hardlines; 24 top Susanna Addario/Thomas Trojer; 26 left Hardlines; 26 top Susanna Addario/Thomas Trojer; 29 Chris Forsey; 30 left Hardlines; 30–31 bottom Chris Forsey; 31 top Susanna Addario/Thomas Trojer; 32 Chris Forsey; 33 Chris Forsey; 36 Susanna Addario/Thomas Trojer; 37 Chris Forsey; 40 left Hardlines; 40 top Susanna Addario/Thomas Trojer; 42 Hardlines; 44 Susanna Addario/Thomas Trojer; 45 Susanna Addario/Thomas Trojer; 46 left Hardlines; 46 top Susanna Addario/ Thomas Trojer; 54 left Hardlines; 54 top Susanna Addario/Thomas Trojer; 62 left Hardlines; 62 top Susanna Addario/ Thomas Trojer; 68 left Hardlines; 68 top Susanna Addario/Thomas Trojer; 72 left Hardlines; 72 top Susanna Addario/ Thomas Trojer; 76 left Hardlines; 76 top Susanna Addario/Thomas Trojer; 78–79 Chris Forsey; 83 all Chris Forsey; 90 bottom left Raymond Turvey; 90 bottom right Chris Forsey; 91 Raymond Turvey; 92 top Chris Forsey; 92 bottom Raymond Turvey; 94 top Raymond Turvey; 94 bottom Raymond Turvey; 95 Raymond Turvey; 98–99 Mark McLellan; 101 Mark McLellan; 104 Raymond Turvey; 106 Raymond Turvey; 110 Susanna Addario/Thomas Trojer; 110–111 Susanna Addario/Thomas Trojer; 115 Mark McLellan; 122 Mark McLellan; 123 Mark McLellan; 124–125 Chris Forsey; 127 Chris Forsey; 130 Raymond Turvey; 131 Raymond Turvey; 140–141 Chris Forsey; 155 Raymond Turvey; 156 top right Raymond Turvey; 156 bottom left Raymond Turvey; 163 Raymond Turvey; 164 Hardlines; 171 bottom Raymond Turvey; 172–173 Wil Tirion; 174–175 Wil Tirion; 176–177 Wil Tirion; 178–179 Wil Tirion;180–181 Wil Tirion; 182–183 Wil Tirion.